10 Easy Turning
for the Smaller Lathe

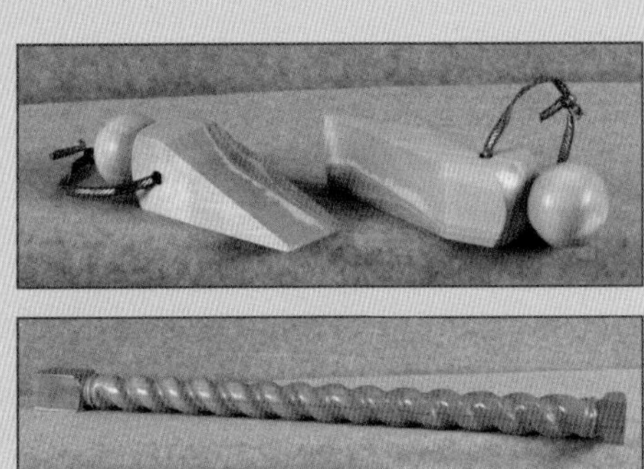

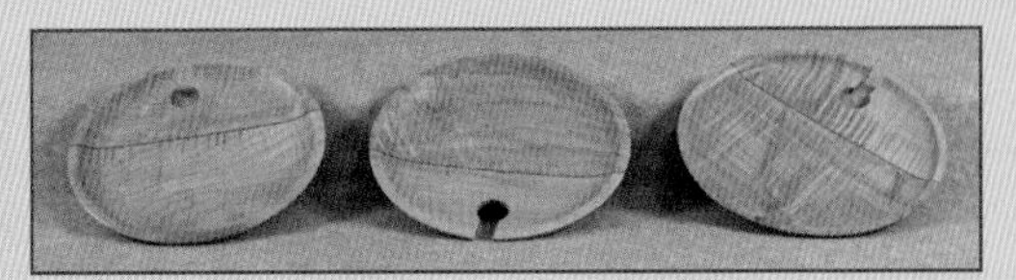

Bill Bowers

4880 Lower Valley Road Atglen, Pennsylvania 19310

Dedication

To Mary K
My only sister whose encouragement over the years added to my many successes in life.

Library of Congress Control Number: 2007933218

Designed by John P. Cheek
Cover Design by Bruce Waters
Type set in Zurich BT

ISBN: 978-0-7643-2727-8
Printed in China

Published by Schiffer Publishing Ltd.
4880 Lower Valley Road
Atglen, PA 19310
Phone: (610) 593-1777; Fax: (610) 593-2002
E-mail: Info@schifferbooks.com

For the largest selection of fine reference books on this and related subjects, please visit our web site at
www.schifferbooks.com
We are always looking for people to write books on new and related subjects. If you have an idea for a book please contact us at the above address.

This book may be purchased from the publisher.
Include $3.95 for shipping.
Please try your bookstore first.
You may write for a free catalog.

In Europe, Schiffer books are distributed by
Bushwood Books
6 Marksbury Ave.
Kew Gardens
Surrey TW9 4JF England
Phone: 44 (0) 20 8392-8585; Fax: 44 (0) 20 8392-9876
E-mail: info@bushwoodbooks.co.uk
Website: www.bushwoodbooks.co.uk
Free postage in the U.K., Europe; air mail at cost.

Contents

Proem

After completing *The Basics of Turning Spirals* my interest was piqued to write a second text on 10 easy projects that could be accomplished on mini- or other smaller lathes. Some of the projects require a bed extension for the mini-lathe but conveniently fit on the old Rockwell, Delta, or other 12-36 inch lathes.

The skill level required is such that beginning to intermediate turners can readily master the tasks involved without angst, trepidation or apprehension. All projects are easily completed on a weekend, yielding a great sense of accomplishment for the neophytic turner.

The impetus, ideas, and intentions for this book come about from my local turning association's beginning classes and club demonstrations. Most members don't have the advanced skill levels to master difficult involved techniques. They do better at schemes which can be completed in a day's time.

To grasp the level of skill required, permit me the indulgence of what our turning classes are and the achievements of the students. There are 5 basic classes. Turning 101 consists of learning how to properly use a roughing gouge, parting tool, skew chisel, and spindle gouge, the appropriate stance, and safety techniques (tying back long hair, removing watches, rings, and jewelry, having no lose fitting clothing especially sleeves, wearing comfortable firm toed shoes, using a face shield and eye protection, dust mask or respirator, and not turning when taking sedative drugs or alcohol). A 3-inch diameter piece of green wood—usually birch or willow in our area of the country—about 7 to 8 inches long is placed between centers and turned to a cylinder with a roughing gouge. The ends are squared with a parting tool, then a skew chisel is used to cut shiny smooth circumferential areas as well as "V" cuts. As frustration levels erupt and intensify, the roughing gouge is used to smooth the stock again into another cylinder, then the spindle gouge is introduced to cut first beads, then coves. All students with access to lathes are encouraged to practice the learned techniques prior to the next class.

Turning 102 is green bowl turning, practicing the proper use of the bowl gouge. Most students are able to complete about 4 to 5 bowls and learn the techniques of cutting bowl blanks out of logs, chain saw safety, and band saw safety as well. Turning 201 is making a candlestick with a base employing the skills learned from the previous classes. Turning 202 is finish bowl turning, power sanding, and vacuum chucking. Efficacy of various types of finishes are discussed and practiced. Turning 301 is making off-centered—oval—tool handles for several high speed 8-inch steel bars—1/4-inch diameter for a small round skew, 1/4-inch diameter for an Eli Avisera small detailing gouge, 1/4-inch square for a rebate tool, and 3/8-inch square for a beading tool. After completion of the 5 classes most individuals have the confidence and competence to accomplish the tasks in this book.

The projects explained here originate from demonstrations given at our association's monthly meetings. There are usually 3 simultaneous demos of different projects keeping onlookers interested even if the demos are repeated in subsequent years. Information on various finishes, staining and coloring techniques, making spirals or applying texturing, and artistic ingenuity are introduced as well.

The stock employed for projects often can be discovered in the bowels of one's shop under a shroud of dust and debris. Scraps from saw mills or fire wood piles yield another source of timber. Nothing seems to go to waste. A visit to the local tree dump may yield exceptional treasures, especially for those who live in parts of the country with outstanding domestic hardwoods. One must be aware of cracks, rot, and checks in stock, and not turn such wood even if filled in with cyanoacrylate (superglue), as the pieces often fly apart at high speed turning.

Since most of the stock tends to be freebies, the small amount invested in non-wood products—bungs, glass inserts, metal oil lamp reservoirs—doesn't require an exorbitant amount of funds.

Start at chapter one and work through the book, trying one project after another and using the skills learned from the previous chapters. Making multiples allows one to have many fine gifts for those special occasions. The repetition produces invaluable practice, experience, and confidence as well. Try modifying projects to personalize the work, but, most of all, have fun and enjoy yourself. By the time one has finished all 10 projects their skills will be honed for more prodigious turning endeavors, for all the techniques learned and practiced will improve one's skill and prowess at more complicated turning accomplishments.

Tools needed are described in the text but most are the standard turning tools with a few exceptions. A small band saw, table saw, jointer, small planer, bench top spindle sander, and drill press are the power tools needed in addition to a lathe. A dust free area is needed to apply finishes.

Chapter 1
Door Stops

The product of a nifty beginning project that can be placed in the home or office is a pair of door stops. All one needs is a 12-inch piece of stock that is somewhere between 2 x 2 inches and 1-1/2 x 1-1/2 inches. The wood needs to be solid without checking and preferably some timber that turns easily. Birch, ash, maple, cherry, walnut or other domestics fit the bill.

For demonstration purposes Alaska birch obtained from saw mill cut-offs (the slabs from canted logs) is cut square on a table saw 2 inches x 2 inches x 12 inches. Mark the center point on each end with a pencil. Various center marking tools are available from turning supply houses (see the first photo in chapter 3).

A steb center in the headstock and live center in the tailstock for the 12 inch piece make the turning easier.

Mount the stock between centers on the center pencil marks.

Bring up the tool rest and adjust the height so that the cutting tip of the roughing gouge is slightly above the horizontal mid-point of the stock—10 to 11 o'clock. Manually rotate the stock to make sure it clears the tool rest before turning on the lathe. The formula for lathe speed is: diameter in inches x RPM = 6,000 to 9000. For a 2-inch diameter piece the speed would be 3,000 rpms. When roughing, however, one should slow down the speed. I usually start at 1500 rpms until the piece is rounded then turn up the speed. Round off about 3 inches on each end. Remember to back a few inches from the end of the stock and work toward the end so that one is roughing from the thicker portion to the thinner portion of the piece (turning downhill with fibers supporting each other as the wood is cut).

Next use a 3/8-inch spindle gouge or 1/2-inch skew chisel to turn spheroids at either end, leaving a nubbin to be cut off on the band saw later. Sand all cut areas using some wax on the sandpaper to cut down dust and heat. Use 100 grit to 320 grit sandpaper as well as steel wool (0000) to finish. If your lathe has a reversible setting, reverse rotations between grits; if your lathe doesn't have a reverse rotation don't worry about it and sand with the work piece spinning on the lathe. Be careful of the square corners when sanding, as small divots may be taken out of one's knuckles. Be very careful with the steel wool as it has the propensity to catch at the nubbins giving one a negative thrill as it madly spins and beats one's hand.

Remove the work piece from the lathe and carefully cut off the nubbins on the band saw—a fine blade such as 12 teeth per inch will give a cleaner cut than rougher blades.

Measure about 1-1/2 inches in from your cut then draw a connecting line with a pencil and straight edge. The line should define equally sized portions.

Saw along the pencil line then cut off the wedge's point to leave it about 1/4-inch thick. Drill a 3/8-inch hole through the side of the stop to be utilized for a loop of rawhide or ribbon so that the stop may be hung on a doorknob.

Finish the pair by sanding all flat surfaces, removing band saw cut marks before applying wipe-on satin polyurethane in a dust free environment. Be careful not to sand the turned areas or flat spots will result.

Chapter 2

Sliding Glass Door Stop

Whenever visiting other people's homes, I've always been disappointed to see a piece of scrap lumber thrown down for a sliding glass door stop. It takes very little effort and time to make a lovely artistic piece that adds that special flare to ordinary scenery. A nice piece of lumber with a double barley twist fits just fine. First one must measure the space, and subtract for the hinge and fastening hook to determine the final length of the stop. After this is done one may proceed.

In this instance a bed extension would be necessary if one has a mini-lathe otherwise the project cannot be accomplished.

As with the standard door stop, measure the center point on both ends of the cherry stock (any left over 2 x 2-inch stock can be utilized in this project, but a softer domestic like cherry works well).

Mount the stock between centers and measure in about 3 inches for the bottom rectangular area.

Measure in about 1 inch for the top square area.

With either a 3/8-inch spindle gouge or 1/2-inch skew, cut a "V" groove at the tailstock end.

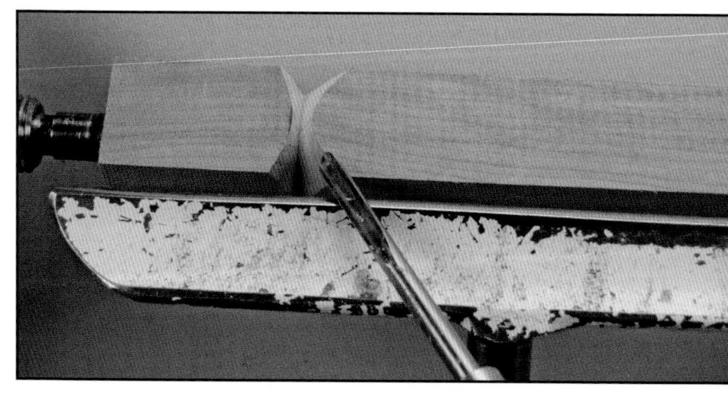

Do the same at the headstock end or bottom along the line.

Next use the roughing gouge to smooth a cylinder between the "V" cuts.

Measure the diameter of the cylinder (1-3/4 inches) to calculate the pitch (degree of slope steepness or grade of the twist). The pitch looks best if it is twice the diameter, that is, 3-1/2 inches.

The completed cylinder is ready for surfacing.

Next draw in the start lines. For a double barley twist there are 2 bines and 2 coves, necessitating drawing 4 start lines—for a more detailed explanation of making twists see *The Basics of Turning Spirals* by Schiffer Publishing Ltd. An indexer on the lathe or a home made indexer is necessary. For a 24-point indexer draw a horizontal start line with a long straight edge at dead center height utilizing the tool rest for support at 6, 12, 18, and 24. If you don't have an indexer divide the measured circumference of the cylinder by 4 and drawn in the start lines.

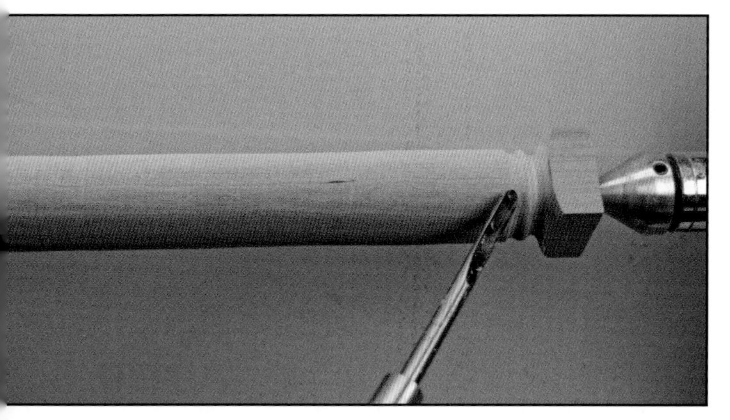

At both tailstock and headstock ends turn a bead and cove for design, allowing the twist to fade into the coves at each end. Make the bead and cove slightly larger at the bottom or headstock end for a more pleasant proportional appearance. Sand the beads and coves before starting the layout for the twists.

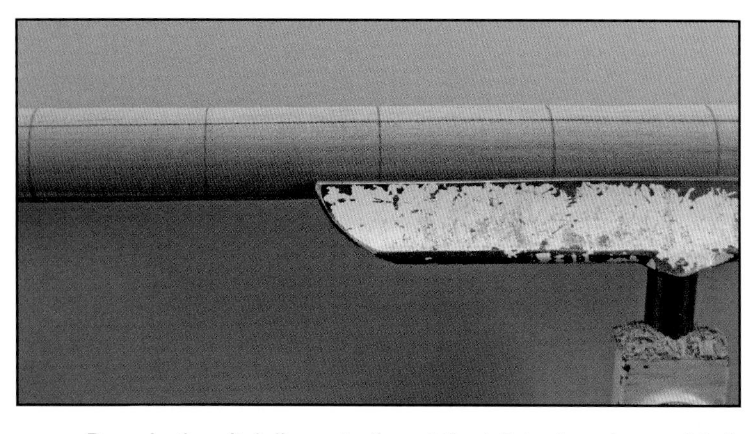

Draw in the pitch lines starting at the tailstock end cove. Mark a circumferential red line every 3.5 inches until the headstock end is reached.

Divide the spaces in half (1.75 inches) by drawing in circumferential red lines.

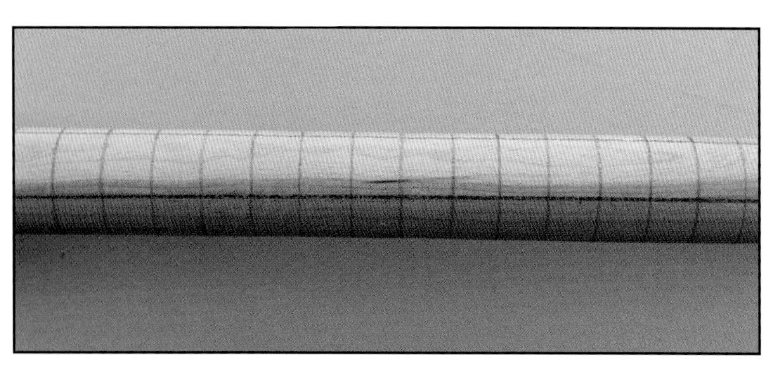

Divide the spaces in half again by drawing in circumferential red lines.

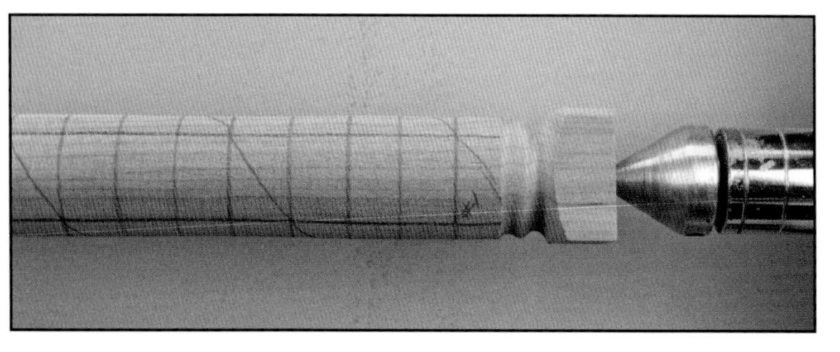

At the tailstock end mark with a blue pencil # 1 in a rectangle. Rotate the cylinder skipping a rectangle and mark # 2. In the rectangle marked #1 draw a diagonal blue line from the lower right hand corner to the upper left hand corner, following into the next diagonal rectangle until the headstock is reached.

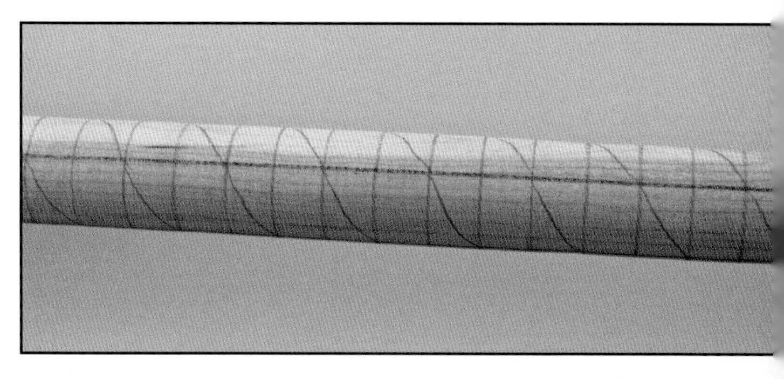

Draw the same type line for rectangle #2. The 2 blue lines are the bine lines for the double barley right handed twist. Had you wished to make a left handed twist you would have started at the headstock end drawing a line from the lower left hand corner to the upper right hand corner of the #1 rectangle around to the tailstock end.

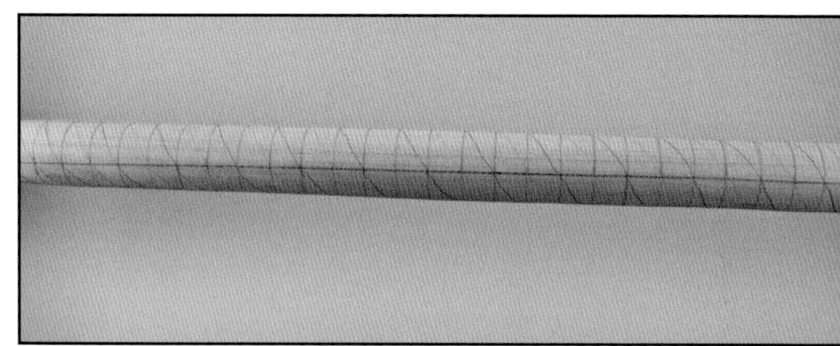

In the two unmarked rectangles at the tailstock draw pink fluorescent lines around to the headstock. These are the cut lines for the double barley twist.

With a dovetail or dowel saw, cut a shallow (1/8-inch) kerf along the pink fluorescent lines. Notice the black tape as a depth gauge on the saw.

Notice both cut lines ready for mircroplaning.

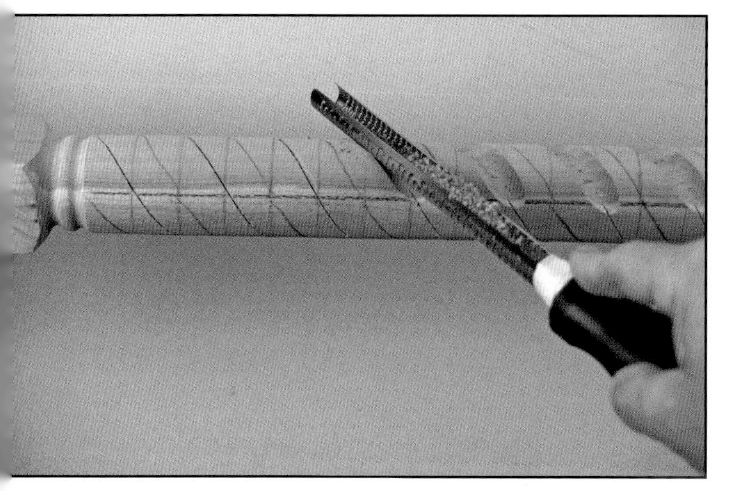

With a large, round microplane begin cutting along the cut lines in a parallel fashion using the saw kerf as a guide. If you wrap some tape around the proximal portion of the microplane it will soften tissue damage effects on the index finger.

Whenever the microplane becomes clogged with shavings tap it lightly perpendicularly on the lathe bed to knock out the debris.

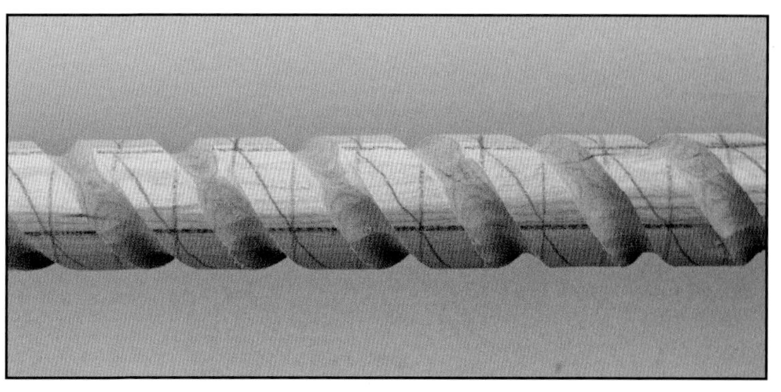

After the coves are partially cut rotate the cylinder continuing the mircroplaning until the two coves are completed. Don't try to remove all the waste material with one rotation but carefully continue the mircroplaning until a continuous consistent cove is created.

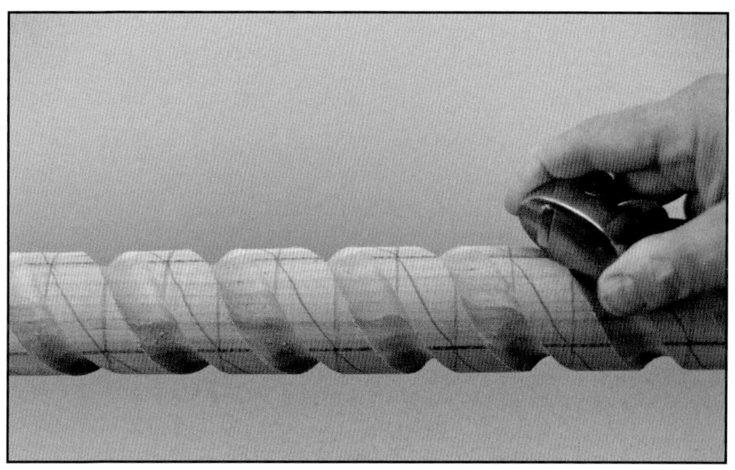

Use a small hobby plane to round over the sharp edges of the bines. I've found the Ibex brass small plane works the best, but it is expensive when compared to the $10 hobby plane.

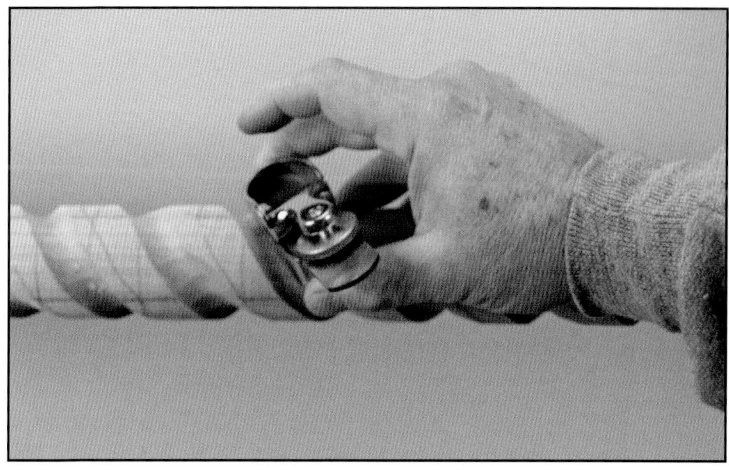

Remember to cut with the plane so as to have the fibers supported otherwise tear-out will occur. For the right-handed, double barley twist, cut the left side of the bines right to left, away from one's self.

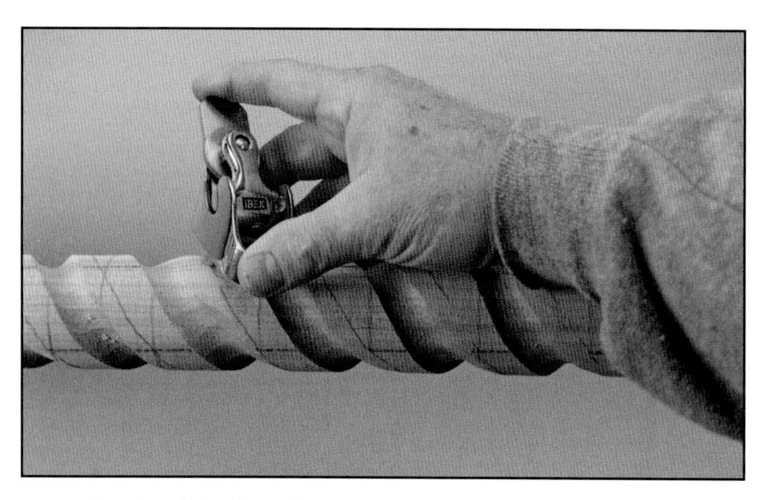

For the right side of the bine cut left to right towards one's self. If you had made a left-handed twist (that means one would have made the initial marks at the headstock and worked to the tailstock end) the cuts would be reversed.

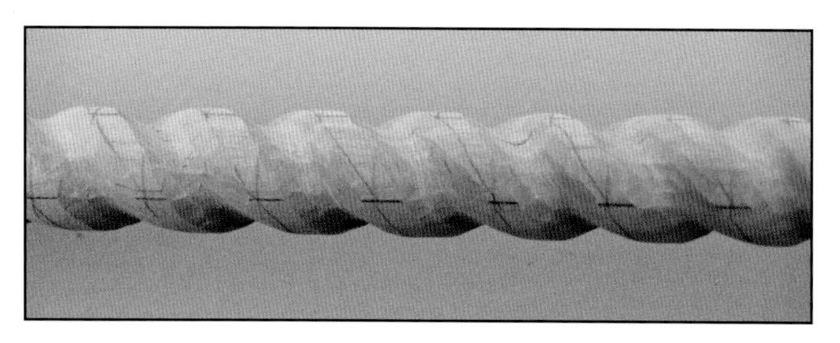

Continue making the cuts until roughed over bines are produced.

Use a 1/2-inch coarse file to smooth over the bines and coves. Remember to file perpendicularly as well as angularly.

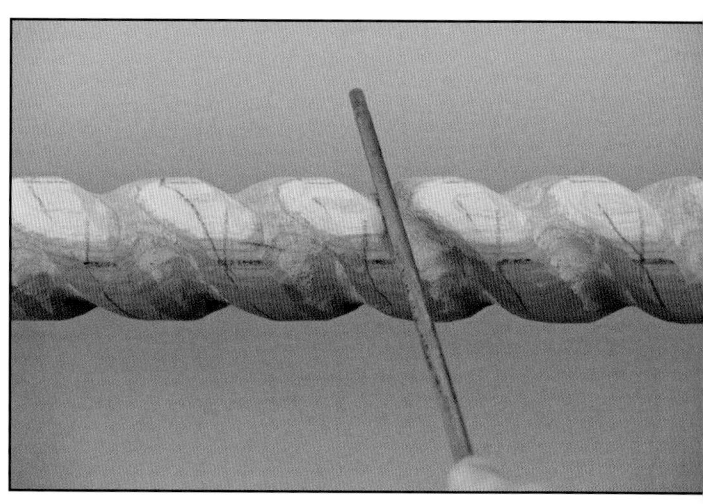

Next use a small fine file to smooth all cut surfaces.

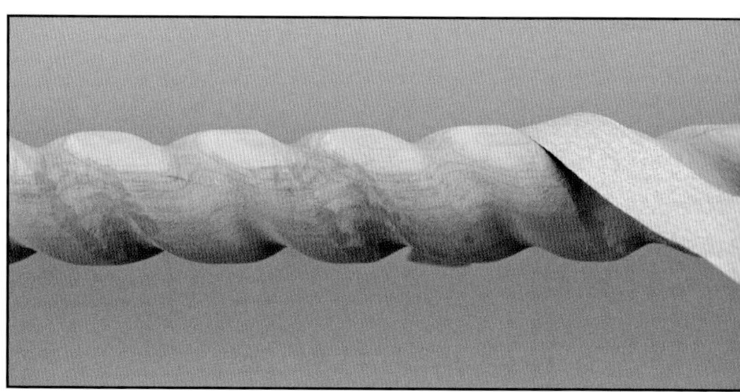

Sand all surfaces with 80 through 320 grit sandpaper. A pliable paper such as Klingspor 1 inch wide yellow rolls or Vitex tearable sheets work best.

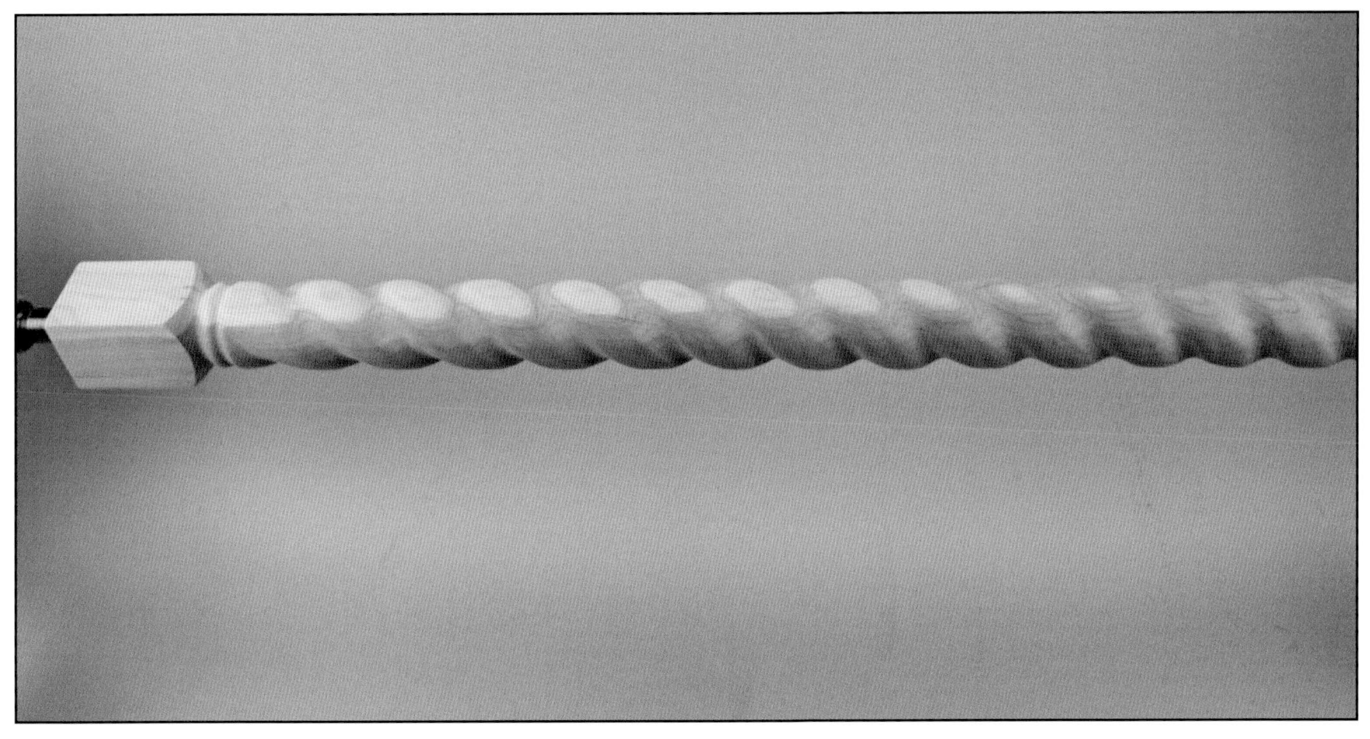

The completed stop is ready for a finish.

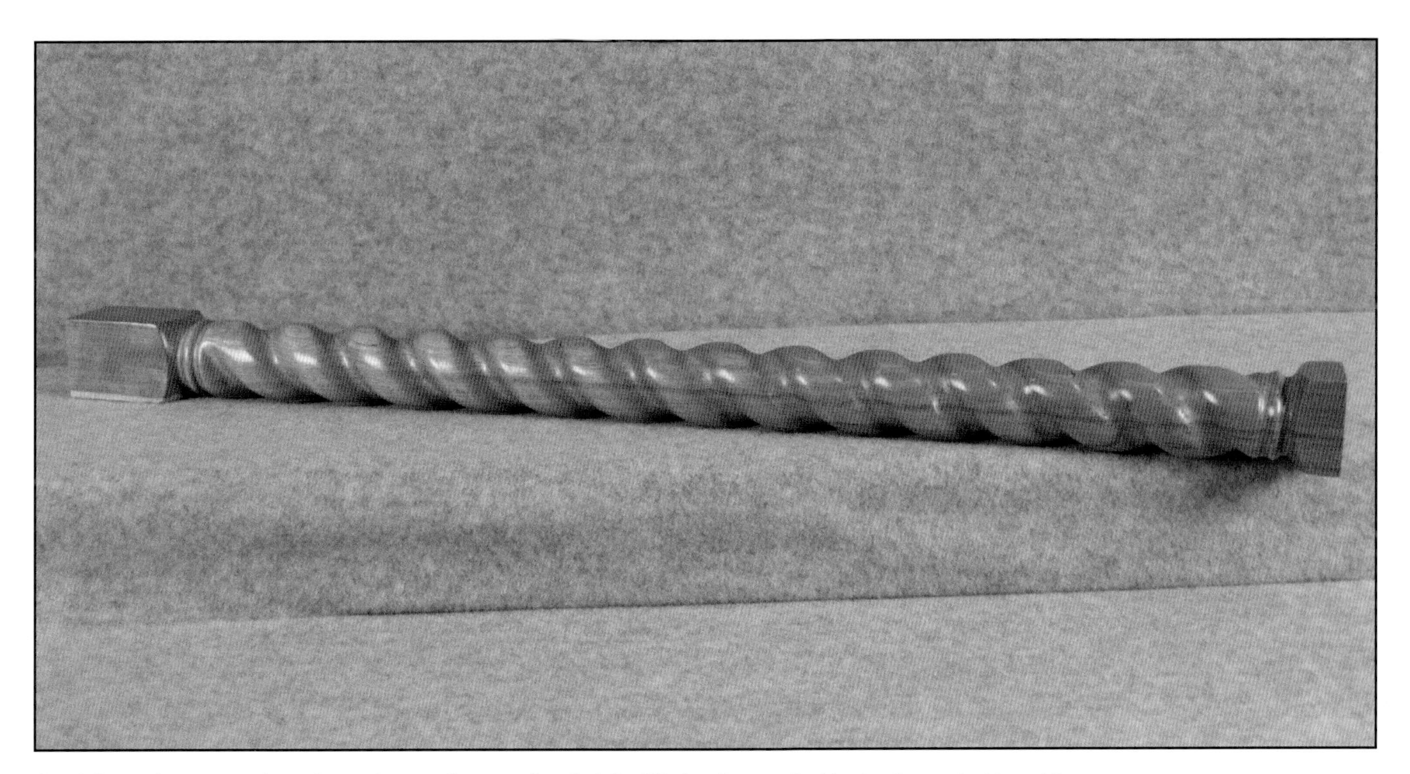

Applying wipe-on satin polyurethane gives a nice finish. All that is needed is the brass bottom hinge and top hook to install the sliding glass door stop.

Chapter 3

Wine Bottle Stoppers

Wine bottle stoppers are neat utilitarian projects that can be made from almost any left over wood, wood products, or other materials. Highly figured wood or burl, especially when stained or dyed, make attractive pieces. The size can vary for the stock from 1-1/2 x 1-1/2 x 2 inches down to 1-inch cubes. As long as there is enough material to glue in place the 1/2-inch dowel a stopper can be made.

I've found that when drilling a 1/2-inch diameter hole 3/4-inch deep, some form of holding mechanism is necessary. One can't safely hand-hold such a small piece as illustrated. With a drill vise it becomes easy, safe, and reliable. An alternative method for irregular pieces would be to place the stock in a 4-jaw chuck on the lathe and drill the dowel hole using a Jacob's chuck in the tailstock. Drilling at low (200) rpms would safely accomplish the task.

Mark the center point and carefully drill a hole 3/4-inch deep in the spalted birch stock—any scrap squared piece (flat, parallel top and bottom) can be used including bone, antler, or synthetics.

Place cyanoacrylate glue on the 1/2-inch diameter long piece of oak dowel to glue it into place. Cut-to-length dowels are commercially available, but cutting your own is much more economical. Make sure you use oak and not poplar, as the latter is soft and will be crushed in the chuck jaws.

A dead blow mallet may be used to pound in the dowel.

After the glue is dry, carefully cut off the four corners so that the stock is easier to turn.

Using a Talon chuck with #1 jaws or equivalent set-up, mount the stock leaving the dowel extending out about 1/2-inch, so that the bottom may be finished easily. Bring up the tailstock with a live center for added stability.

Turn the stock to a cylinder with a 3/8-inch spindle gouge or 3/4-inch roughing gouge.

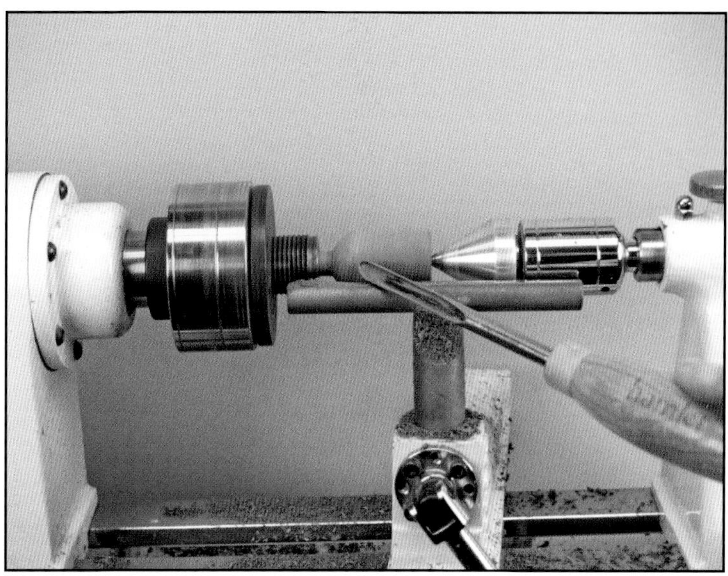

Round over the bottom with the spindle gouge or 1/2-inch skew chisel.

Cut a broad concavity in the body of the piece.

Smooth out the rough spots, then back off the tailstock.

Carefully cut away the central nubbin off the top and form a subtle concavity.

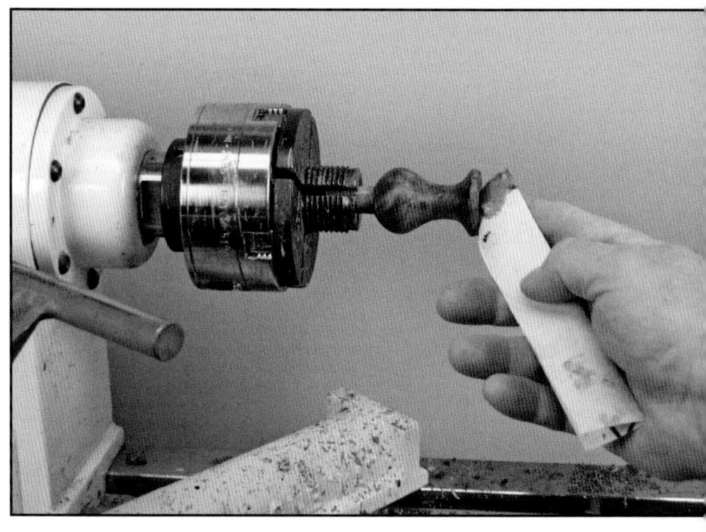

Sand all cut surfaces with waxed sandpaper using 100 to 320 grit, finishing with 0000 steel wool. Be careful not to get the steel wool caught in the chuck jaws or wrapped around the dowel.

An interesting effect can be had by applying water based gel stains with a brush to lighter or figured woods—spalted Alaska birch in this instance.

Allow several minutes for the stain to dry—this particular product is ready for a finish after 1 hour drying time.

Wipe off excess stain with a paper towel or cloth.

Apply another color (purple) to the bright red for a burgundy coloring.

After several minutes wipe off the excess stain.

Utilizing left over pieces often yields interesting bottle stoppers. Here black palm cut-offs give a nice short bottle stopper—good for placing left over white wine in refrigerators.

After the stopper is dry apply at least 2 coats of wipe-on satin polyurethane—non-stained stoppers require only one. Since the stain is water-based, there is always some lifting of the grain. This requires the stopper to be remounted and buffed again with 0000 steel wool.

The finished product is quite nice.

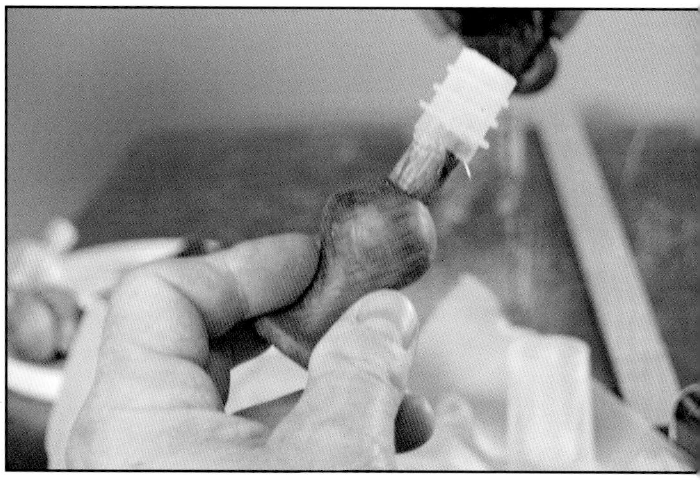

Place a dollop of silicon sealant on the dowel before twisting on the bung—a small tube of silicon sealant is good for about 400 bottle stoppers.

Occasionally the dowel is cut too long or the drilled hole is too shallow, necessitating trimming of the dowel for the silicon bung to fit properly. This can be accomplished on the band saw using a fine-toothed blade and wood slat brace; otherwise one may have a nasty accident.

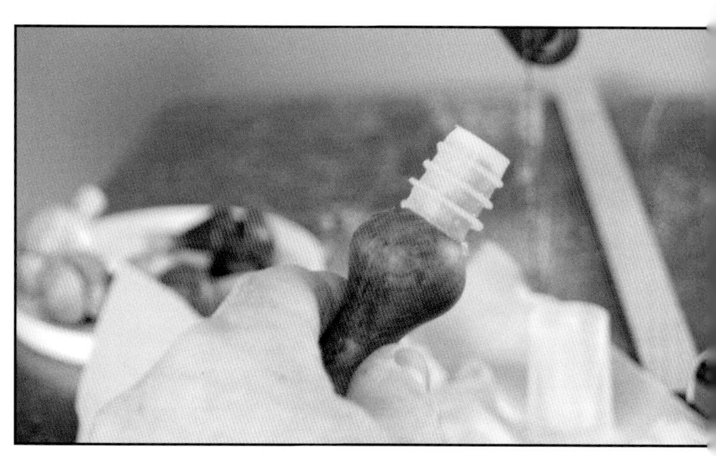

Allow the finished bottle stopper's silicon sealant to dry about one day before use. The reason for the glue is to keep the bung from dislodging into the wine bottle when the stopper is removed. The fit with the silicon bung is so good that a full 750 ml bottle may be lifted by the stopper without coming loose.

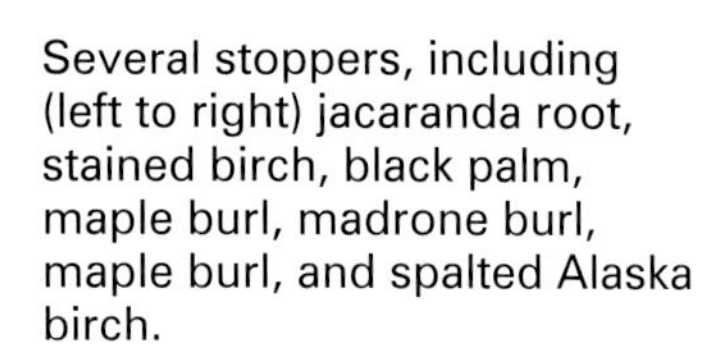

Several stoppers, including (left to right) jacaranda root, stained birch, black palm, maple burl, madrone burl, maple burl, and spalted Alaska birch.

Chapter 4

Confetti Oil Lamps

Making confetti oil lights, vases, or lamps is another easy project where waste wood finds a function. A chunk about 3 inches thick cut to a 4-inch diameter will do just fine. Usually cut-offs from burl bowl blanks will yield the appropriate stock, but sometimes glue-ups from left-over projects or other figured wood will also fit the purpose. The stock for this project comes from a waste maple burl pile at my friend Burt's shop, where I often visit and turn for fun.

Try to have a semi-flat surface for the top and bottom and cut the diameter as round as possible on the band saw. At the center point (where the compass point mark is) drill a 1/4-inch hole 3/4 inches deep for the screw chuck.

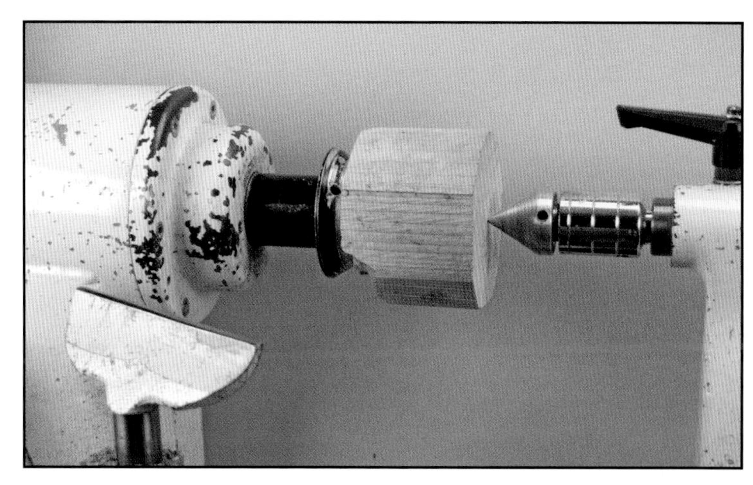

Screw on the stock so that there is no space between the chuck and stock, then bring up the tailstock for support.

There are many commercial screw chucks available. On the far left is the Oneway standard screw chuck, which fits into the Talon or other Oneway jaws. It is a bit large in diameter for our projects but does well for bowls. In the center is another screw chuck requiring a 1/4-inch diameter, 3/4-inch drilled hole. The chuck on the right is my favorite, a Glaser screw chuck, which will be used throughout the various projects. Unfortunately, as with most great products, it is no longer available. If you are fortunate enough to own one, great; if not, get the center chuck from one of the turning supply houses or use the Oneway screw.

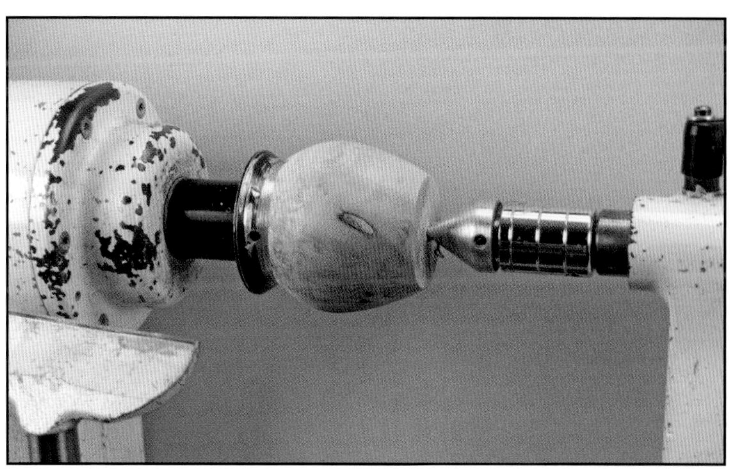

Turn a vase shape with a 1/2- or 3/8-inch spindle gouge.

With calipers mark a 2-inch diameter area on the bottom to fit the 2-inch O'Donnell jaws. Remember to touch only one of the points to make the mark. If both points touch the calipers will be violently withdrawn from one's hand leaving a bent pair of calipers.

Fashion a foot by turning a subtle concavity and mark several design lines with the tip of a skew chisel. Using a 3/8-inch spindle gouge, taper the lines of the vase to the foot. Sand the foot and foot area with 100-320 grit waxed sandpaper, then use 0000 steel wool.

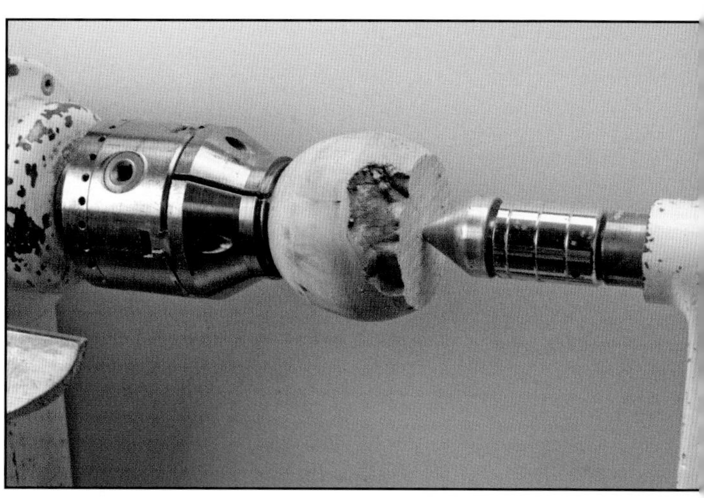

Mount the foot in the 2-inch O'Donnell jaws with a firm, but not tight, grip. If the jaws are tightened too much, marks will be left on the foot. Bring up the tailstock and continue to shape the vase.

Using the 3/8-inch spindle gouge smooth over the curve to the live center.

Back off the tailstock and remove the remaining waste wood. Next, using a 1-1/2-inch diameter Forstner bit drill a 1-1/2-inch deep hole.

Drill the hole at slow rpms (200) and use a little beeswax on the bit for a lubricant.

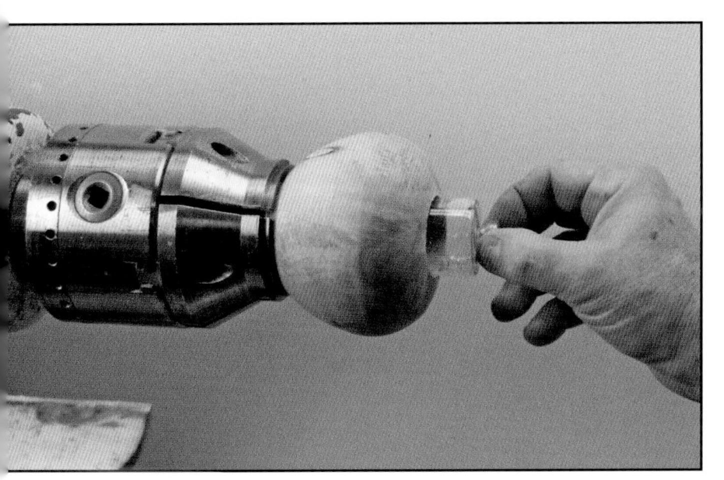

After blowing out any shavings from the drilled hole, check the fit of the glass insert to make sure it is properly seated.

Continue fashioning a lip about the drilled hole with the 3/8-inch spindle gouge then sand the surfaces with waxed sandpaper 100-320 grits. Finish with 0000 steel wool. Apply wipe-on satin polyurethane as a finish.

After the 2 coats of wipe-on polyurethane are dry, remount the confetti light using the 1-inch O'Donnell jaws in expansion mode so that the few lumps and bumps from adherent dust may be buffed off with 0000 steel wool. (Unless one is applying finishes in an operating room with positive pressure atmosphere, there will always be bits of dust adherent to the finish product no matter how careful one is.)

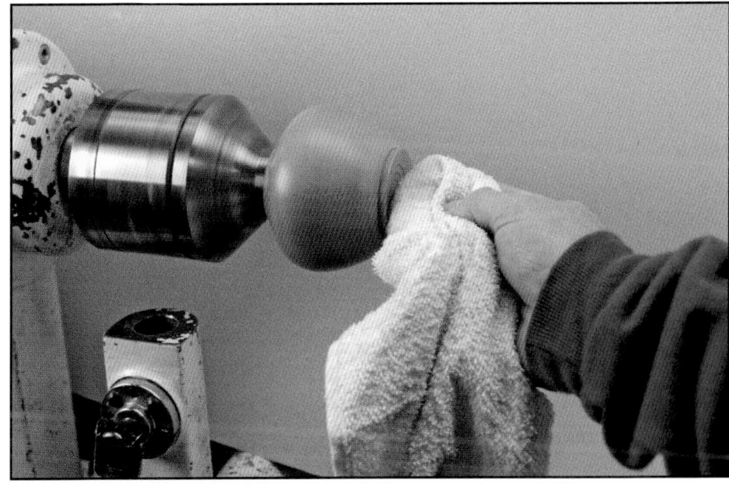

After buffing with steel wool apply a paste wax such as Briwax.

Buff off the wax with a clean cloth.

Insert the glass oil reservoir and fiber-glass wick for the finished product.

If one is making several confetti lamps for gifts, don't place the wick as most people pick up the lamp and turn it over to view the bottom allowing the glass encased wick to fall to the floor and fracture. Also, suggest to recipients of the lamps use of liquid paraffin as a fuel because it is odorless and smokeless, unlike the colored oils sold with the glass inserts.

Chapter 5

Oil Lamps

Another easy project is constructing a standard oil lamp, once again from left over lumber. A good start is a piece about 6 inches square and approximately 4 inches thick. Preferably wood with figure or some left over exotic looks best. There are several metal inserts commercially available for use in the lamp. They usually come with a standard glass flue. I happen to like the Lamplight product which comes in pewter or copper finishes.

A 6 x 6 x 4-inch block of myrtle is cut to a disc on the band saw.

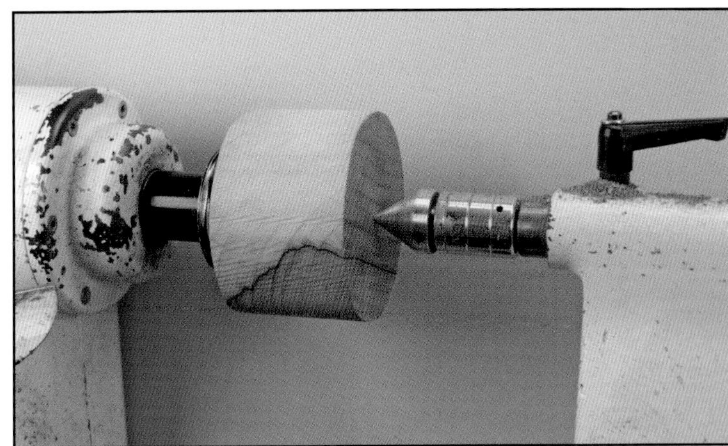

Mount the disc on the Glaser chuck, then bring up the tail-stock for support.

After cutting the disc, drill a 1/4-inch, 3/4 inches deep hole to fit the Glaser chuck.

With a 1/2-inch spindle gouge, turn a vase shape.

Mark a 2-1/2-inch diameter with calipers for mounting. Remember to touch only one of the points of the calipers when making a mark.

Using a 3/8-inch square skew or beading tool, cut a 1/4-inch deep dovetail recess along the caliper mark.

Back off the tailstock and, using a 3/8-inch spindle gouge, cut a concave to convex recess making sure the center of the recess is below the outside foot area.

Use the 3/8-inch spindle gouge to roll several beads.

Sand the bottom and recess with 100-320 grit waxed sandpaper and finish with 0000 steel wool.

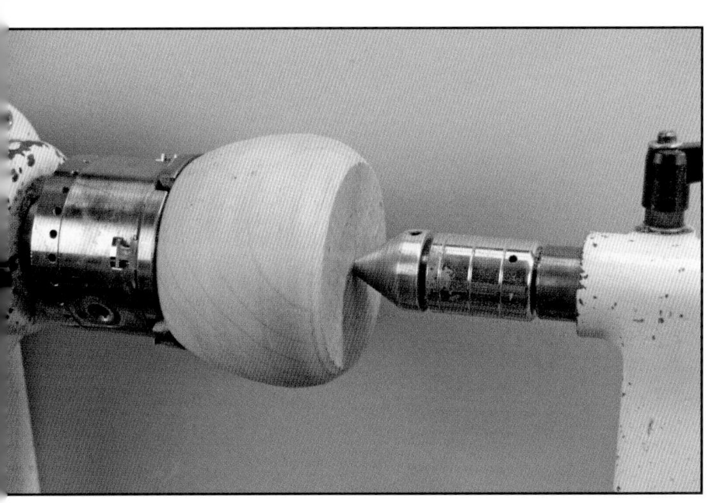

Mount the lamp in the Axminster dovetail expanding jaws or equivalent chuck and bring up the tailstock. Continue to shape the top portion of the lamp with the 3/8-inch spindle gouge.

Mark the diameter of the lamp insert reservoir with calipers after squaring the top surface.

Finish smoothing out any defects or "woolies" of end grain with a sharp 3/8-inch spindle gouge.

Use a sharp bowl gouge with a 42-50 degree cutting angle to hollow out the lamp's bowl.

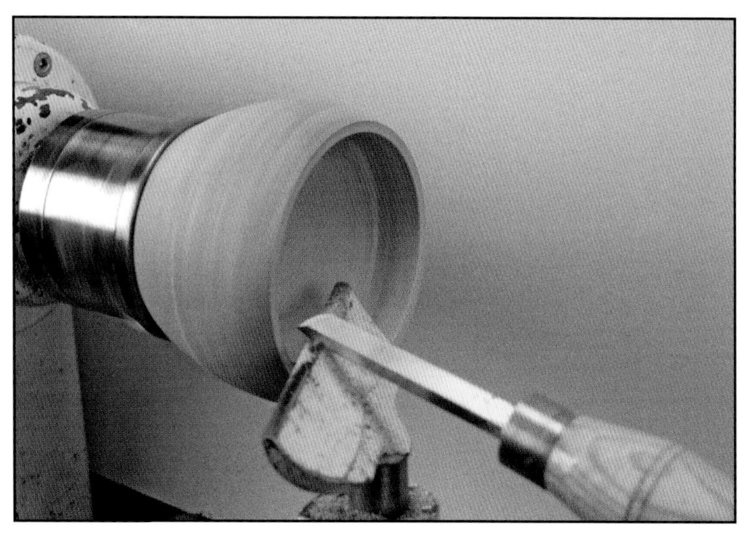

With the 3/8-inch square skew, cut straight sides so that the metal reservoir fits. Check the fit before removing the lamp. Complete by sanding all visible surfaces (100-320 grit waxed sandpaper and 0000 steel wool), then apply 2 coats of wipe-on satin polyurethane in a clean finishing area.

After the finish is dry remount the lamp base and lightly buff with 0000 steel wool to remove any marks or defects.

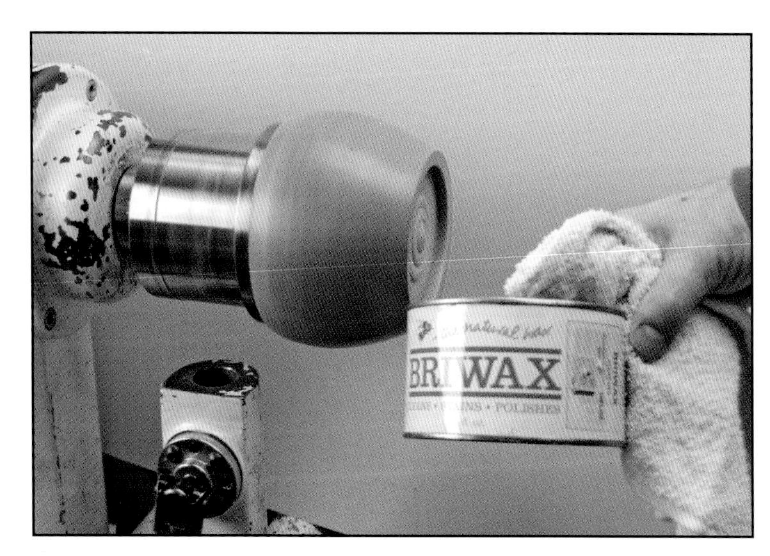

Apply a paste wax.

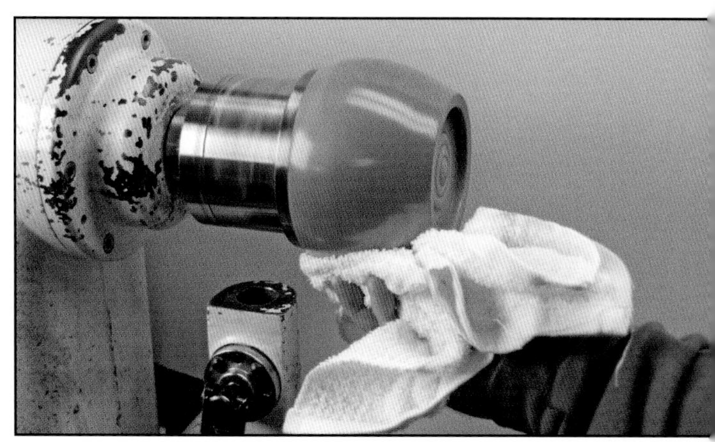

Buff with a clean cloth.

Remove the lamp base and insert the copper finish reservoir. A dab of silicone sealant in the inside base and along the upper border will glue in the reservoir. As with the confetti lamps use liquid paraffin for fuel.

Chapter 6

Candlesticks

Making candlesticks with a base combines spindle with cross grain turning. It is a project used in our turning association's 201 classes and also in the classes our turning association teaches at local high schools. The timber utilized is usually cut-offs from canted logs.

To cut discs properly, use some 1/4-inch plywood discs 6 inches in diameter (it is a good idea to have such discs handy, graduated by 1/2-inch diameters up to whatever the maximum size one's lathe can handle. The discs are used in cutting blanks with one uneven side—this is especially good for cutting bowl blanks). Use a short screw through the center to fix the disc to the stock.

Screw the base onto the Glaser chuck as tightly as possible without stripping the drilled hole. Turn the base square.

Cut out the base by carefully cutting along the attached round disc then drill a 1/4-inch diameter hole 3/4 inches deep using the drill press.

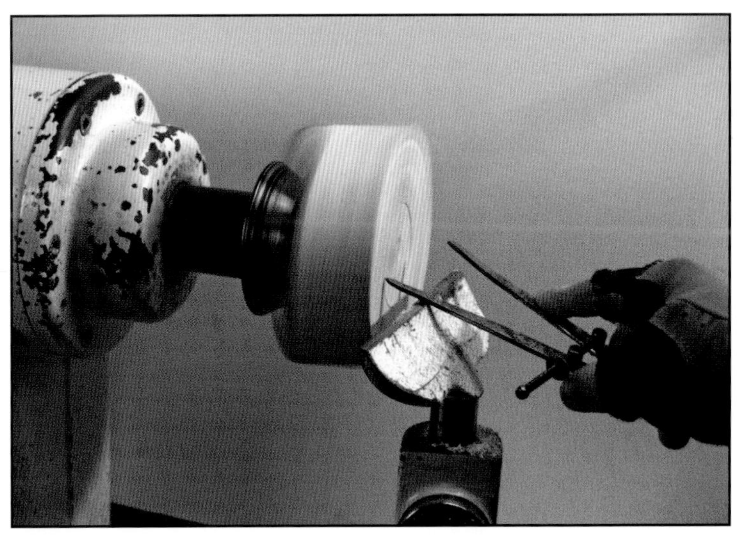

With calipers mark a 2-1/4-inch diameter to fit the #2 Talon jaws in expansion mode or equivalent chuck.

Turn a 1/4-inch deep recess as well as several beads. Sand the bottom and recess with waxed sandpaper 100-320 grit before using 0000 steel wool.

Mount the base on the Talon chuck.

After turning off the bark and creating a smooth flat surface about 1.5 inch in diameter in the center drill a 1/2-inch hole about 7/8 inches deep. Drill at 200 rpms and lubricate the bit with beeswax.

Turn a large ogee form on the base and sand with waxed 100-320 grit sandpaper, before using 0000 steel wool. Remember to form the ogee by cutting from the flat surface towards the headstock—turning down hill—so that the fibers are supported upon each other.

Place a 3-inch long piece of 1-inch diameter dowel in the Oneway or equivalent jaws and taper the distal one inch to 3/4 inches. This will be the mandrel to hold the candlestick spindle.

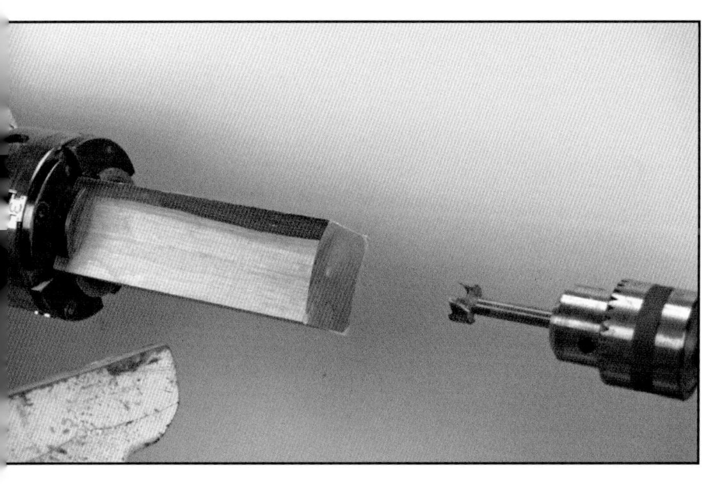

Mount a squared 2 x 2 x 7-inch piece of stock in the Oneway jaws and carefully turn a slight concavity with the 3/8-inch spindle gouge at 3000 rpms. If one becomes too aggressive in the turning, the stock will be knocked out of the jaws. A light touch with a razor sharp tool is most helpful.

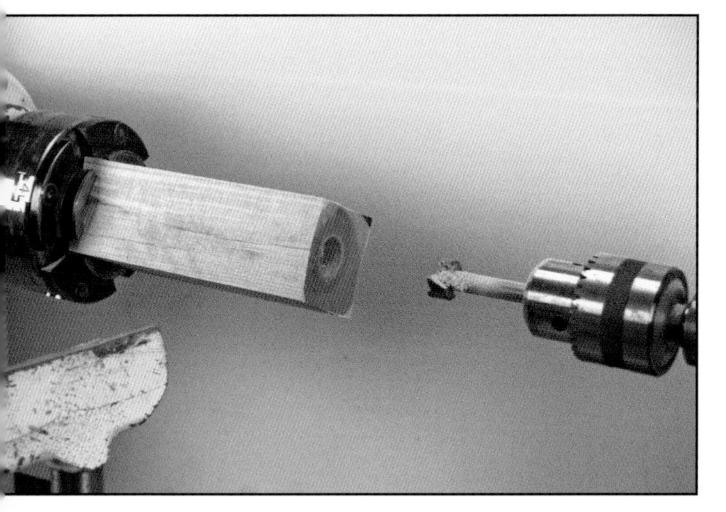

Using a 7/8-inch Forstner bit, drill a 3/4-inch deep hole at 200 rpms. Lubricate the drill bit with beeswax.

Reverse chuck the stock and drill a 1/2-inch diameter hole 7/8 inches deep. This will receive the 1/2-inch diameter 1-1/2-inch long dowel that will fasten the spindle to the base.

Mount the stock between centers using the previously turned wooden mandrel. Make sure the mandrel is square by bringing up the tailstock to the previous point mark before tightening the jaws. Use a roughing gouge to turn a cylinder. Remember to roll the roughing gouge towards the chuck end so that the jaws aren't touched with the gouge. If one becomes too aggressive in turning the cylinder, the stock will merely spin rather than fracture.

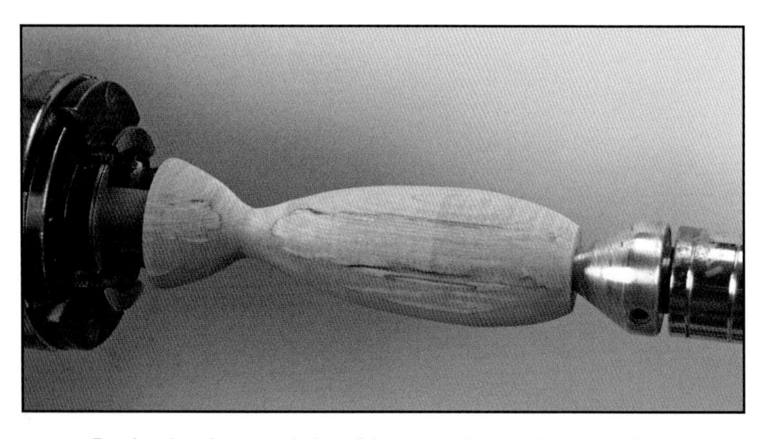

Begin shaping a subtle wide-curved cove towards the top and a half parabola to finish the bottom.

Glue the spindle to the base with the 1/2-inch dowel then finish with wipe-on satin polyurethane. After the finish is dry buff with 0000 steel wool to remove bumps.

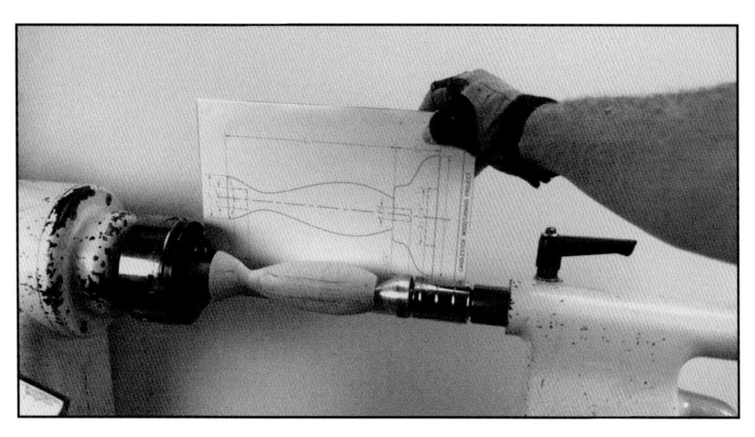

For our classes we have a life-size pattern for the students to check the correct shape. After the spindle is turned sand with waxed sandpaper (100-320 grits), then use 0000 steel wool.

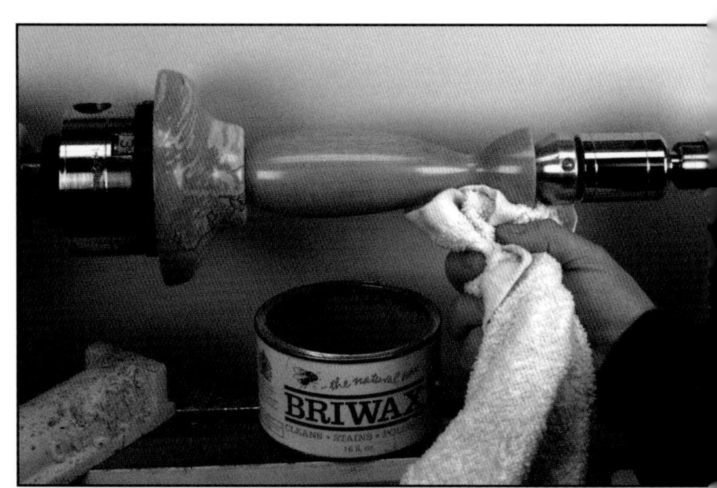

Apply a paste wax, then buff with a clean cloth.

The finished pair of candlesticks will make a lovely gift. Had one wanted to paint the pair in an Art Deco style, no wax would have been used in sanding and, of course, no polyurethane would have been applied.

Chapter 7

Wine Caddies

Wine caddies are wonderful, practical gifts to keep red wine stains from tablecloths, fine finished wood, or laced placemats. They are another utilitarian, easily made object. The stock used is something 1-1/2 to 2 inches thick and about 6 inches in diameter. Cut-offs from other projects are usually available to produce them. Old pieces of lumber with checks or cracks can be used for timber after the defects are cut away. The piece demonstrated comes from my friend Arnie's wood pile and is partially spalted with many cracks and punky inclusions. Wine caddies can be made plain or have their rims stained for accents.

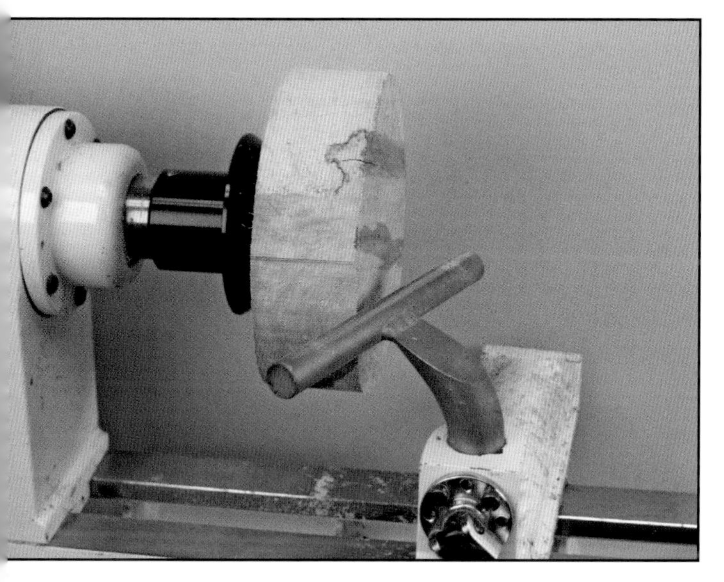

The 6-inch diameter 2-inch thick Alaskan birch stock is mounted on the Glaser screw chuck for shaping.

The completed ogee form and squared bottom is ready to have its rim rounded.

With a 52 degree bevel on the 1/2-inch bowl gouge, use push cuts to shape the ogee form.

Round the rim with the same push cut from either side of the rim to prevent tear out.

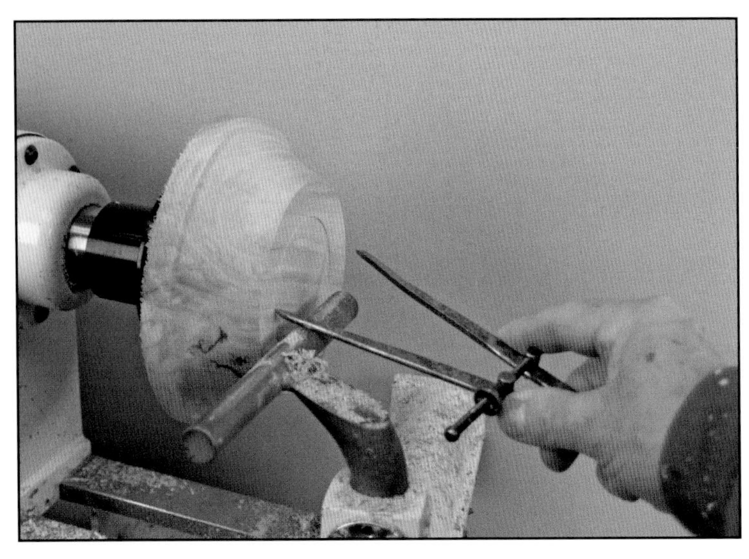

With calipers, mark a 2-1/4-inch diameter area to fit the Talon #2 jaws or equivalent chuck.

Using the 3/8-inch spindle gouge roll several beads for design then sand the bottom to completion.

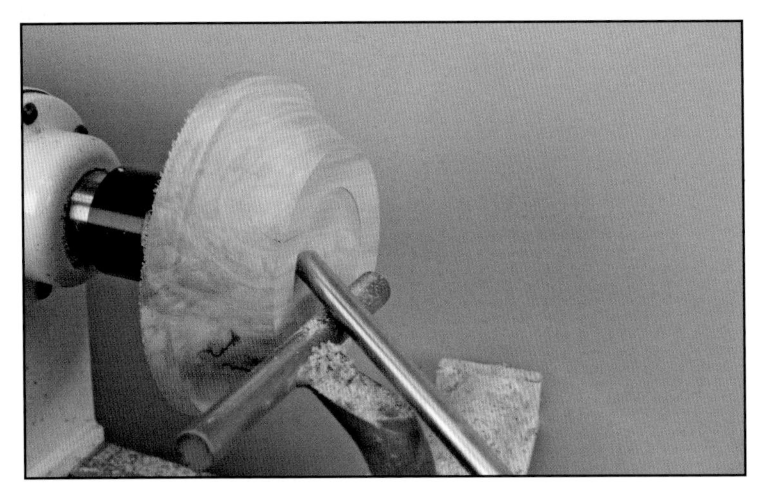

Use the bowl gouge to cut in towards the center a concavity to convexity with the high point lower than the caddy's rimmed foot.

Use the Talon chuck with #2 jaws for the mounting.

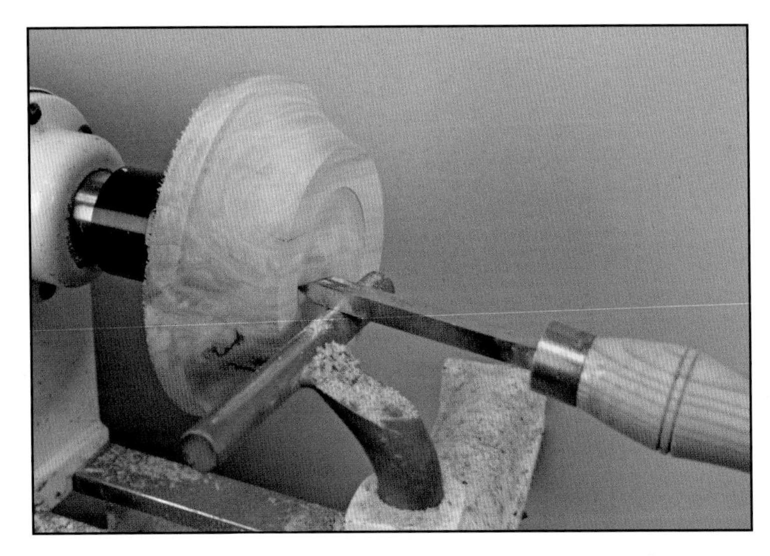

With the 3/8-inch skew cut a 1/4-inch deep dovetail for the Talon jaws.

Mount the caddy by expanding the jaws slightly into the dovetail. Don't apply a death grip or the soft wood will be fractured.

Begin turning off the thick rim to one of about 3/16-inch, tapering towards the center.

Turn out the center mass of the caddy by starting in the center and working back to the rim just like one does for bowls. The passage from the side to and across the bottom should be in one continuous motion rotating the gouge tip to float on the bevel as the center point is reached. The initial cuts should have the flute closed—opened area parallel to the lathe bed. As the bowl gouge is floated on its bevel towards the center, gradually open the flute so that when the center is reached its opening is perpendicular to the lathe bed.

Notice the 52 degree angle cut on the bowl gouge.

Use the 3/8-inch skew to place design lines in the base and on the rim. Sand the piece to completion before applying wipe-on satin polyurethane finish.

If one wishes to apply stain as a design element it should be done after the backside is polyurethaned with two coats and the rim front surface is finished and sanded (without wax of course). Brush on the desired water based stain and allow it to dry for several minutes. Wipe off the excess then cut in several design lines with the 3/8 inch skew tip.

Sand the cut lines then apply with a very thin brush a contrasting stain in the grooves.

After the stain is almost dry wipe off the excess with a clean cloth. Finish hollowing out the wine caddy.

Apply 2 coats of wipe-on satin polyurethane before remounting.

Use 0000 steel wool to knock off small defects such as adherent dust.

Apply a paste wax then buff with a clean cloth.

The finished wine caddies present an interesting use of throw away lumber.

Chapter 8
Prep Plates and Small Bowls

Prep bowls and plates are another easy object to produce from left over lumber. The idea comes from a friend who is a gourmet cook. He wanted about eight different small plates and bowls for holding spices, condiments, and flavor accents when preparing his gourmet meals. Somehow many of the prep bowls and plates I have made appear around my home with paperclips, candy, stick pins, or pills in their respective reservoirs, making it obvious that they have many functions beyond their use in cooking. The plates and bowls are usually 6 inches in diameter or smaller, finished with walnut oil if the wood is non-toxic or with polyurethane if made from exotics or woods with high tannic acid content.

The Glaser screw chuck is once again utilized but a 1/4-inch thick spacer is needed because the stock is not thick enough to drill a 3/4-inch deep hole.

After cutting a 6-inch disc of the Honduran red heart stock on the band saw, drill the proper depth to fit the chuck.

Measure the extension of the screw so that an appropriate sized 1/4-inch hole can be drilled in the stock.

Mount the stock and turn the side true and the base square.

Mark a 2-inch diameter foot with calipers.

Using a bowl gouge (52 degree bevel) with the flute closed, use push cuts to round over the plate.

Notice the 52 degree bevel on the bowl gouge.

Use a 3/8-inch spindle gouge to cut the 1/8-inch deep foot.

With the same gouge, roll several beads on the foot.

Sand the outside surfaces to completion with 100-320 grit waxed sandpaper, then use 0000 steel wool.

Mount the plate by its foot in the 2-inch O'Donnell or equivalent jaws with an easy grip. If you tighten the jaws too much marks will be left on the foot requiring another procedure to finish the bottom. Use a 1/2-inch bowl gouge with a 52 degree bevel to hollow out the plate. Cut with the flute closed, leaving a center support. Next use a bowl gouge with a 72 degree double bevel to smooth the center flat and make a smooth transition with the previous cuts. Sand surfaces with waxed sandpaper and 0000 steel wool. Apply wipe-on satin polyurethane.

After the plate is mounted in the O'Donnell jaws, it is hollowed out in the same fashion as the red heart.

Sometimes left over glue-ups can make interesting plates. The stock demonstrated is Alaska birch cut-offs from cutting boards. After cutting the stock round on the band saw, it is mounted like the red heart plate was and turned true. A foot is turned for mounting in the O'Donnell jaws.

After the plate is turned to completion, finish with sanding as before, then apply a finish.

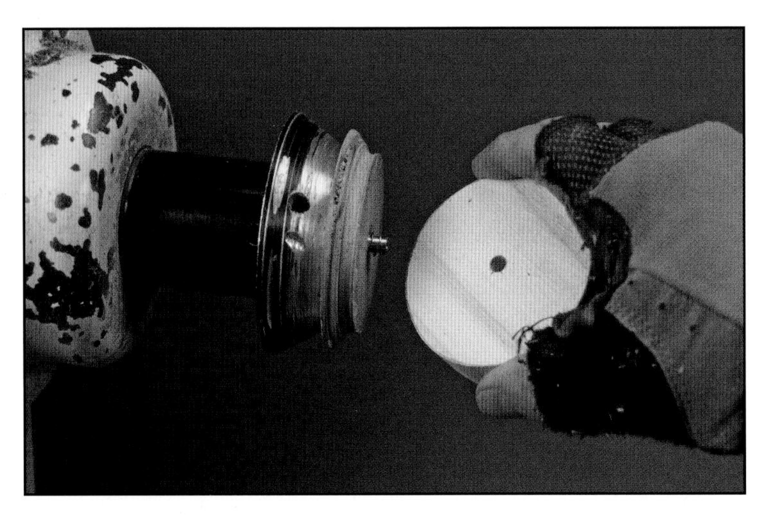

Small left-over pieces of wood, like this 3/4-inch thick piece of ash cut to 3 inches in diameter, make interesting small bowls. A thicker spacer is needed to turn the stock and a shallower drill hole is needed.

The stock is mounted, rim trued-up, and base squared.

Mark a 1-1/4-inch diameter area with calipers for a foot, then turn a bowl shape using push cuts. Turn the foot to 1/8-inch thick and place some lines for design on the bottom with the skew chisel.

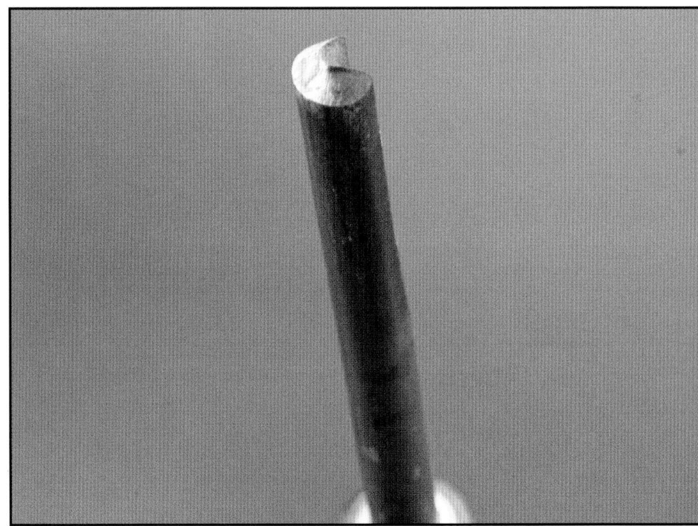

Use a 72 degree bevel to hollow out the bowl as there is a subtle concavity to the bowl.

Whenever making the finishing cuts start at the inside rim with the flute closed and with fluid motion gradually open it as the center is approached.

One can readily see that a smooth cut is obtained and no scrapers are necessary. Sand to completion and apply a finish.

Canary wood is a pleasant timber to use for small bowls. The 2-inch thick stock here has been cut to a diameter of 5 inches then drilled.

There is no need for spacers with this thick stock. Turn a bowl shape, foot, and place a design on the bottom.

Mount the foot in O'Donnell jaws and turn the inside of the bowl leaving some stock in the center for support.

Turn out the bottom mass of the bowl with the 72 degree beveled bowl gouge. Use an opened flute technique then sand to completion. Apply a finish.

Some very thin or shallow plates can be turned, foot placed, and mounted for finish turning.

As with the small bowls or bottoms of the thicker bowls, use the 72 degree bevel to smooth out the interior before sanding and applying finishes.

By removing the chuck, the knot hole defect is readily discernable and approachable for repair. Note: It is best to leave the stock mounted as thin pieces may be impossible to remount in the same special relationships as before.

Some bowls with defects can be turned and repaired to make usable pieces.

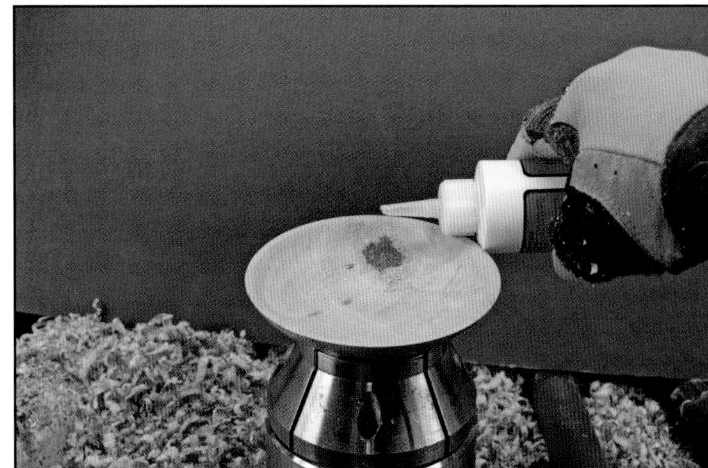

Apply a few drops of cyanoacrylate thin then hand sand the area with 80 grit paper to make acceptable filler before placing the chuck back on the lathe and completing the sanding and finishing.

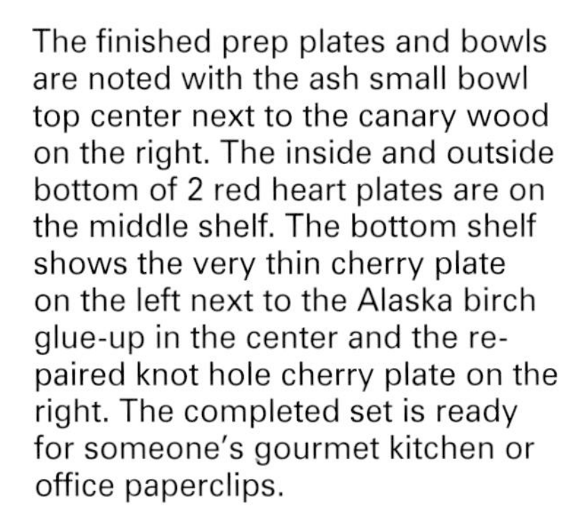

The finished prep plates and bowls are noted with the ash small bowl top center next to the canary wood on the right. The inside and outside bottom of 2 red heart plates are on the middle shelf. The bottom shelf shows the very thin cherry plate on the left next to the Alaska birch glue-up in the center and the repaired knot hole cherry plate on the right. The completed set is ready for someone's gourmet kitchen or office paperclips.

Chapter 9
Hors d'Oeuvre Plates

The same friend who wanted prep plates for his kitchen activities also wanted hors d'oeuvre plates that would hold a wine or champagne glass. Most hors d'oeuvre plates are 8-½ to 9 inches in diameter and are easily turned on the smaller lathe.

The project can utilize many different types of left-over lumber with a thickness of about 1 inch. One factor to consider, however, is to use only solid pieces that are quarter sawn, otherwise the plate will warp in spite of its small size. I find gluing several pieces together with some contrasting veneer can make attractive plates and utilize a lot of left over small pieces of lumber, but some people like plates without those accents. As with the wine stoppers and wine caddies, a stain may be applied for accents. One does need to use a jointer to square the surfaces to be glued. A glue that takes stain or finishes is mandatory. Polyurethane glue works great especially if you wear latex gloves in its application (mainly to keep the sticky glue off one's hands) and utilize good clamping techniques.

After cutting out the 9-inch diameter disc measure the distance the Glaser screw extends beyond the spacer disc to get the depth of the 1/4-inch hole to be drilled.

Use a 1/4-inch disc spacer on the Glaser chuck with a lesser diameter than the curly maple disc so that the edges may be turned without fouling on the spacer.

Drill the appropriate hole in the disc's center—the compass point mark used to draw the 9-inch circle.

Mount the disc by turning it on so that no gap is left between the stock and the spacer.

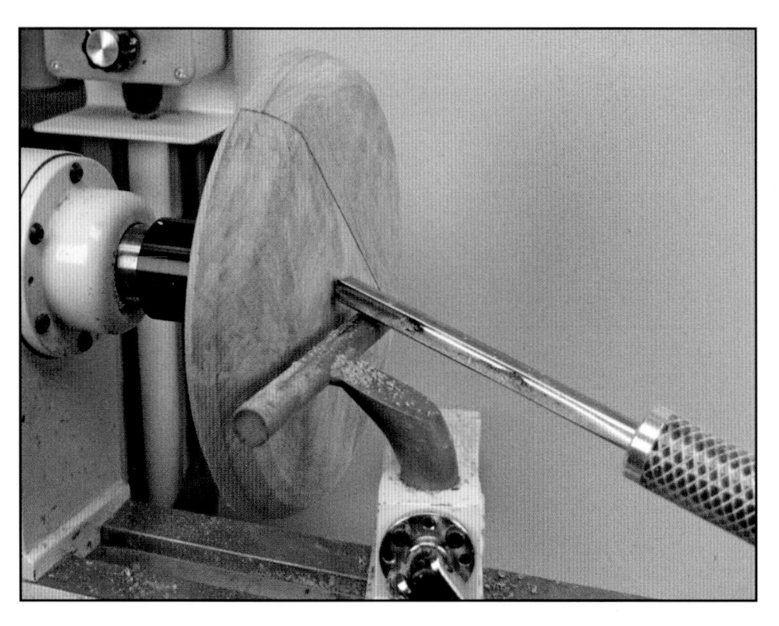

Square the bottom with a 1/2-inch bowl gouge.

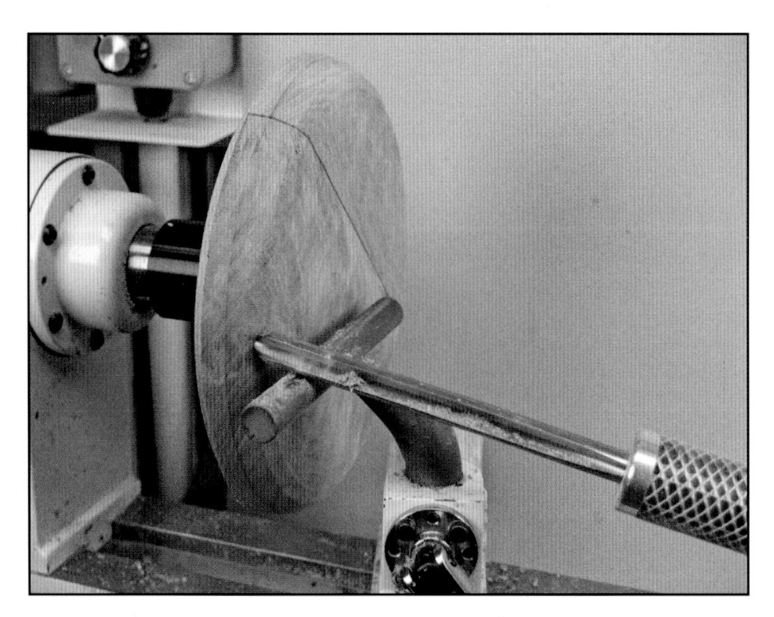

True the edge of the plate and turn a slight round over.

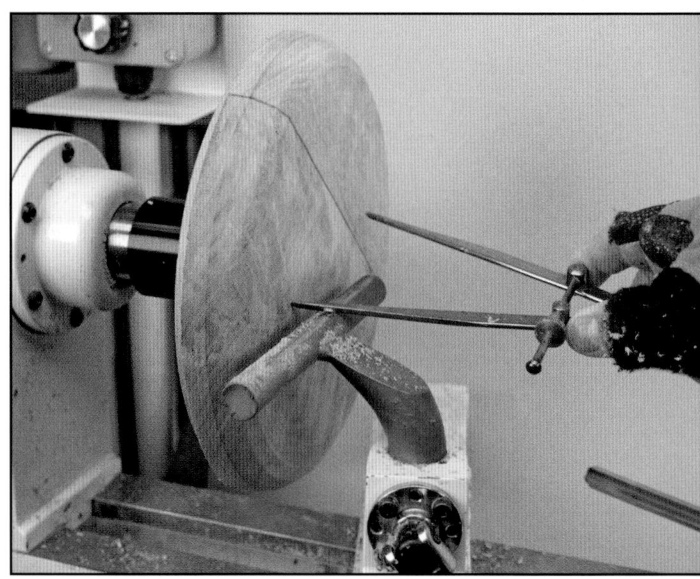

Measure a 4-inch diameter area to fit the Axminster 4-inch jaws or equivalently sized chuck.

Make sure you touch only one point of the calipers to make the mark.

Using the 1/2-inch bowl gouge, begin cutting a flat surface from the caliper mark to the plate's curve.

Use the 3/8-inch spindle gouge to cut the 1/8-inch foot.

Use the 3/8-inch spindle gouge to smooth over the plate's curve.

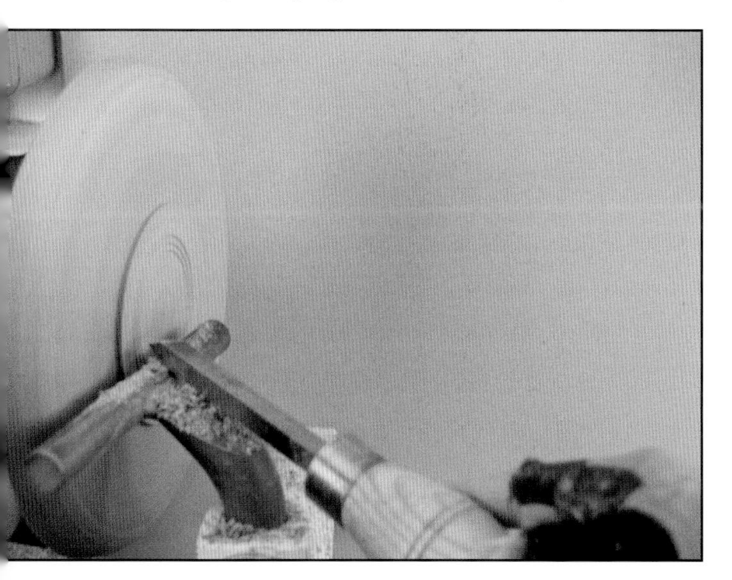

With the 3/8-inch skew's point mark several design lines on the foot.

Sand the bottom to completion.

Next mount the plate in the 4-inch Axminster dovetail jaws.

Remember to apply a firm but not tight grip so that there will be no marks left on the foot when it is completed and removed.

Begin cutting the contour of the plate, starting from the lip and working in towards the center. Use the gouge with the 52 degree bevel for this shallow plate.

Leave some mass in the center when shaping the curve of the plate so that the support stabilizes peripheral vibrations.

Turn out the center mass with the 72 degree beveled gouge so that the plate's thickness is about 1/4 to 3/16 inch. Remember to have a subtle, nearly non-discernable concavity in the center. This is for liquid material that may accompany the various hors d'oeuvres. If the plate is left flat the liquid may run out the opening cut for the wine glass.

Sand to completion and buff with 0000 steel wool.

The next step is cutting the hole for the wine glass. A 1-inch hole made with a Forstner bit will be large enough for 99.9% of the glass stems. Mark a point 2 inches in from the lip and support the lip on that side with a slat. This will prevent the plate from tipping whenever the drill bit is applied to the uneven surface.

Flip the plate over and use the small hole as a quide for the Forstner bit point to drill though.

Drill down about 1/8 inch or until a small hole is seen on the reverse side.

By drilling from either side, blowout of the wood fibers is prevented.

Next cut a 1/2-inch path to the drilled hole on the band saw with the plate's face down.

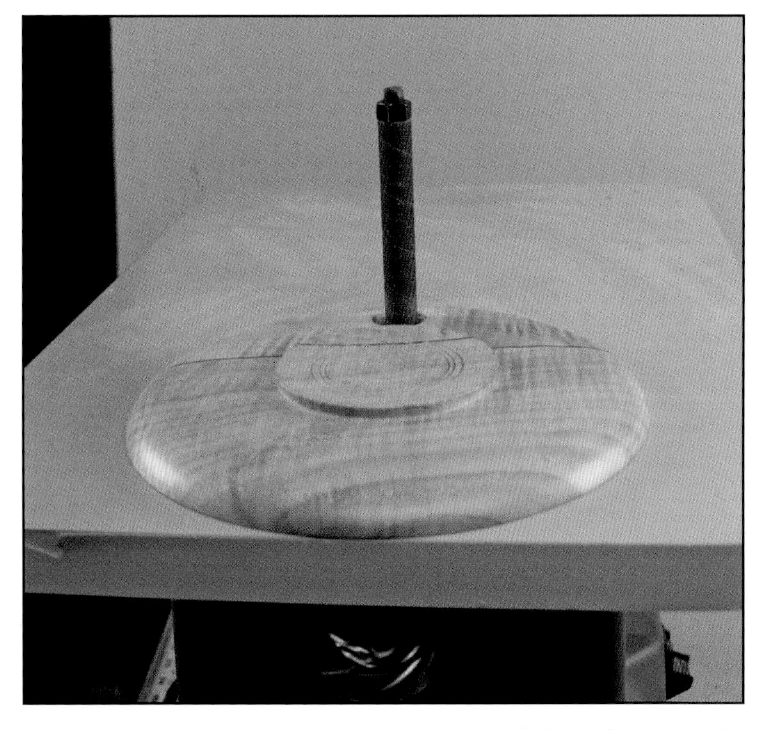

Using a table top spindle sander and a 1/2-inch sleeve, sand the cut opening and the drilled hole smooth.

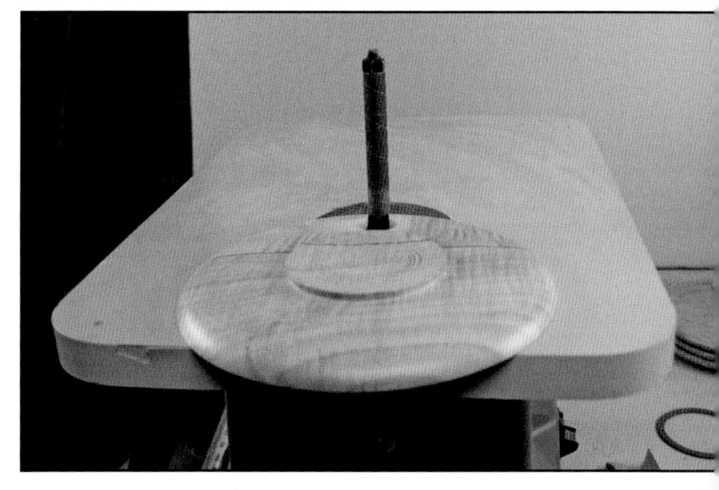

Going back and forth will sand off any band saw marks.

The finished plate is ready for polyurethane to be applied. I usually apply at least 3 coats to each side separately, since food will be placed on the plate and it will be cleaned off with a soapy cloth. More protection and a thicker finish is necessary than with other products.

Buff both sides of the plate with steel wool but do it free hand. If you mount the plate and use steel wool the drilled and cut hole will tear apart the steel wool.

Several of the hors d'oeuvre plates made from glue-up curly maple and purple heart veneer are ready for action.

One can readily see how handy the plate functions. Holding the plate with wine glass in place allows one to have a free hand to eat, gesture, or shake hands while pontificating political problems.

Chapter 10
Plant Stands

Plant or large candle stands make nice projects. They can be produced simply or have various complicated embellishments added to make an interesting center or conversation piece. Since I happen to have a lot of cherry left over from previous projects it is used for the plant stands.

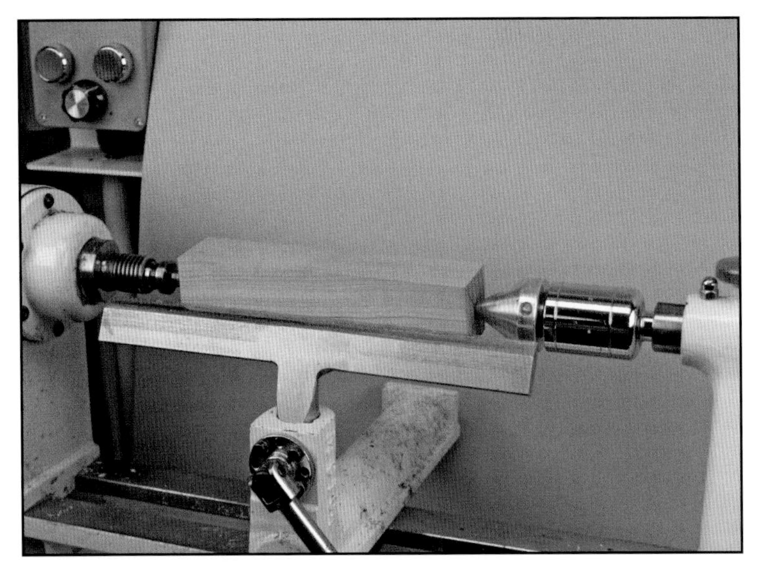

Mount a squared 2 x 2 x 7-inch piece of cherry between centers.

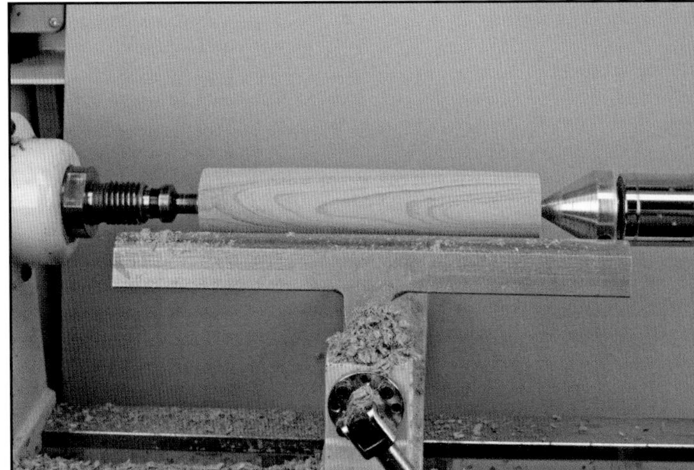

Form a smooth cylinder.

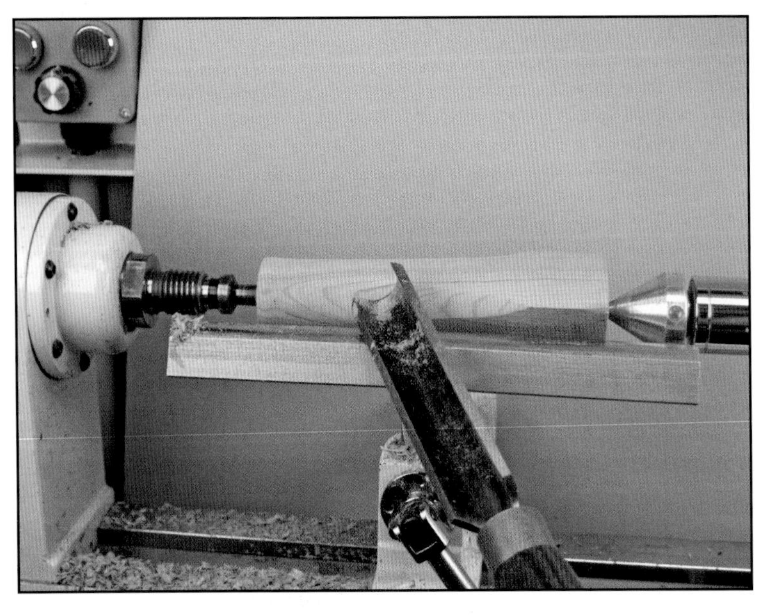

With a roughing gouge, begin turning a cylinder, starting from one edge and working backward so that one is turning downhill.

Next, using a broad parting tool, cut a spigot (3/4 inches diameter, 3/4 inches long) to fit into the top and base of the stand.

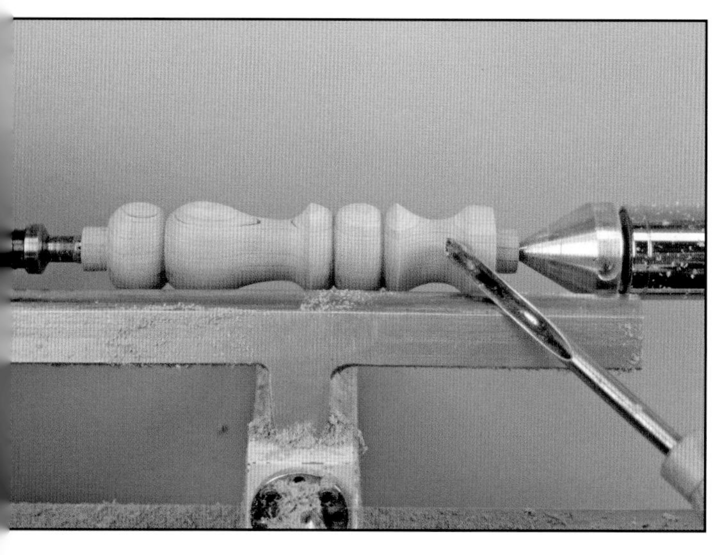

Turn an attractive design on the spindle with a 3/8-inch spindle gouge.

Cut out a 4-1/2-inch diameter, 2-inch thick piece of cherry. Drill a 1/4-inch diameter, 3/4-inch deep hole at the compass point mark then mount the stock on the Glaser screw chuck.

Sand to completion and finish with 0000 steel wool.

Measure the outside opening of the #2 Talon jaws with calipers.

Turn the cherry base true and square the sides.

Mark the 2-1/4-inch diameter on the bottom.

With a 3/8-inch square skew cut a 1/4-inch deep dovetail.

From the dovetail cut, using the 3/8 inch spindle gouge, cut a concavity to convexity making sure the high point is deeper than the rim. Roll several beads with the spindle gouge for design.

Sand the base and recess to completion.

Next mount a 2-inch thick, 7-inch diameter piece of cherry on the Glaser screw chuck and round over the underside.

Sand the piece to completion.

Square the face with the spindle gouge cutting in about 1/4-inch.

Mount the base using the Talon chuck and #2 jaws in expansion mode.

Round over the base to make a 1-3/4-inch diameter flat surface.

Using a 3/4-inch Forstner bit, drill a 3/4-inch deep hole to fit the bottom spigot of the plant stand spindle.

Check the fit to make sure there is no gap between the spindle and base.

Sand to completion.

Mount the plant stand top in the Oneway flat jaws with rubber bungs holding the finished piece. Drill a 3/4-inch deep, 3/4-inch diameter hole to fit the top spigot of the spindle. Glue the three pieces together and apply a finish. Note: If one doesn't have the flat jaws, a jam chuck can be fashioned out of plywood or MDF to fit the plant stand top. However, if one plans to do much turning, invest in the necessary equipment, as it can be easily used over and over again.

An interesting stem can be made for a plant stand by utilizing an inside-outside turning. An attractive embellishment to be added could be a twist in the center. Using left-over cherry, run it through a planer until a perfect 1-1/2-inch square is produced. Cut off 6 inch sticks and number them 1 through 4.

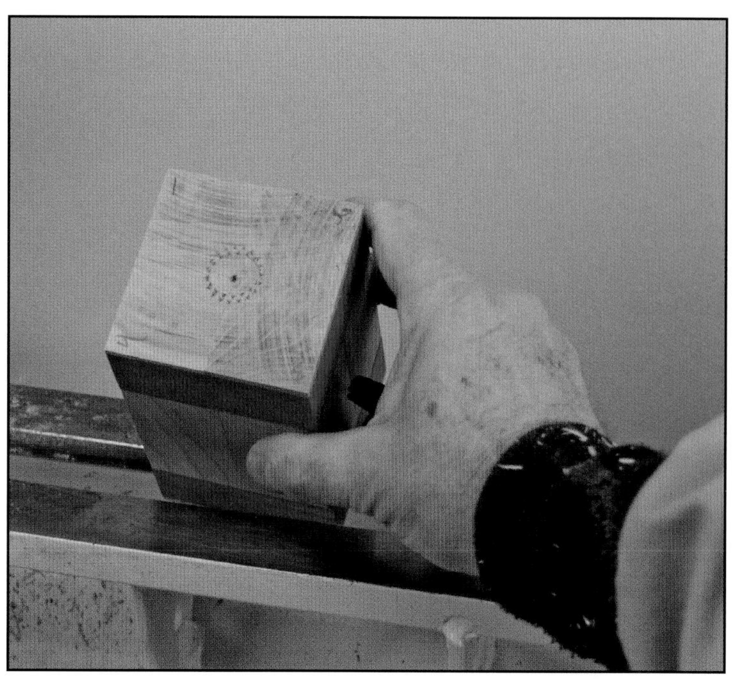

After squaring the ends on a chop saw, place the taped piece between centers, using a steb center with its center point and a cup center with a center point. Tighten the tailstock so that marks are made in the wood. Remove the stock and the center point from the steb center and cup center.

Tape the numbered squares together firmly.

Remount the piece using the steb center and cup center marks for guides. Screw in the tailstock cup so that a firm grip is accomplished. The cup center has its outside tapered inward so that it holds together the four pieces quite nicely. The center points are removed because they tend to push apart the taped wooden pieces.

Next turn a pleasing cove and sand it to completion. Apply a water based gel stain and let it dry for a few minutes. Wipe off the excess with a clean cloth.

Undo the tape and lay out the turned pieces.

Rewrap the pieces and check the numbering to make sure the pieces have been rotated 180 degrees before gluing. Retape the glued pieces and let dry about a day.

Mount the stock in the #3 Talon jaws or equivalent chuck after marking the center points for the steb and cup centers. Drill a 1/2-inch hole through the top for placement of the center twist.

Reverse chuck the piece and drill a hole through the bottom.

Check the fit of the steb center before remounting the stock. The steb center should be dead center.

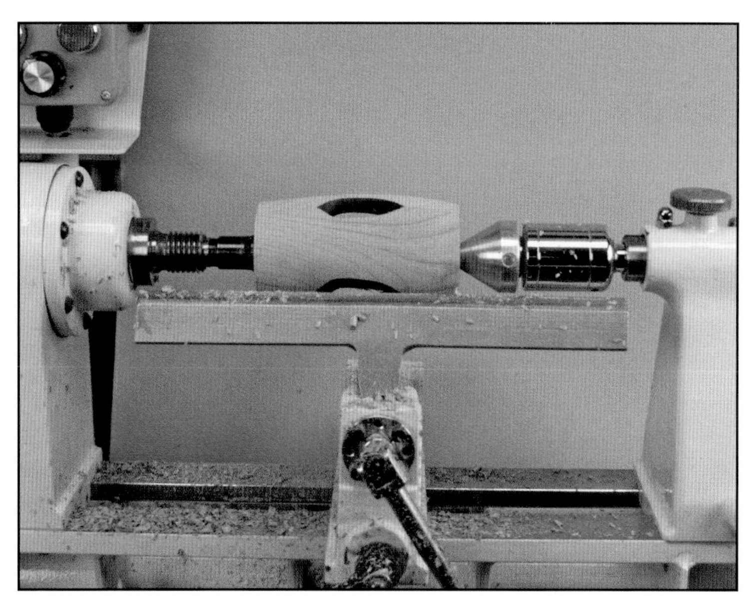
Turn an oval pattern to open the 4 windows in the piece.

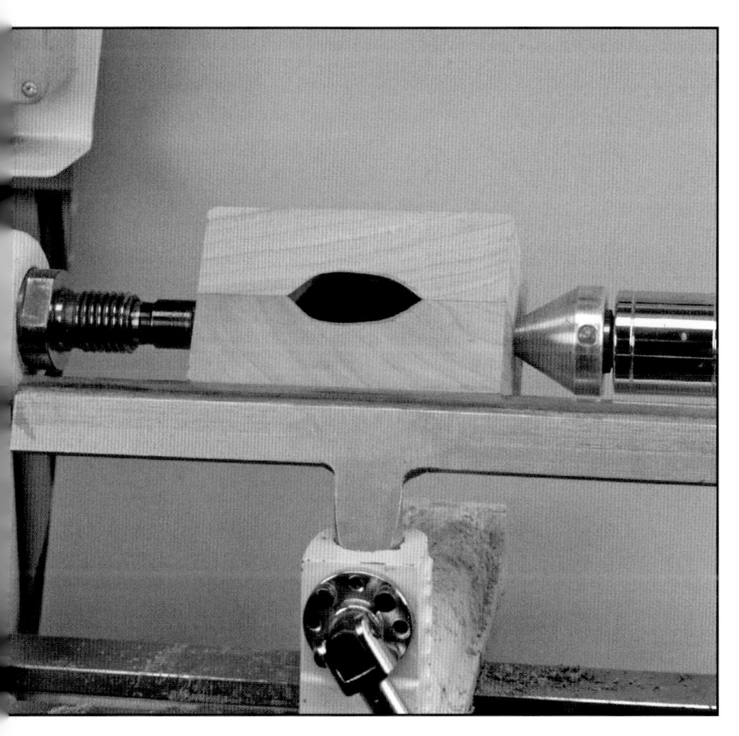
Mount the stock and remove the tape.

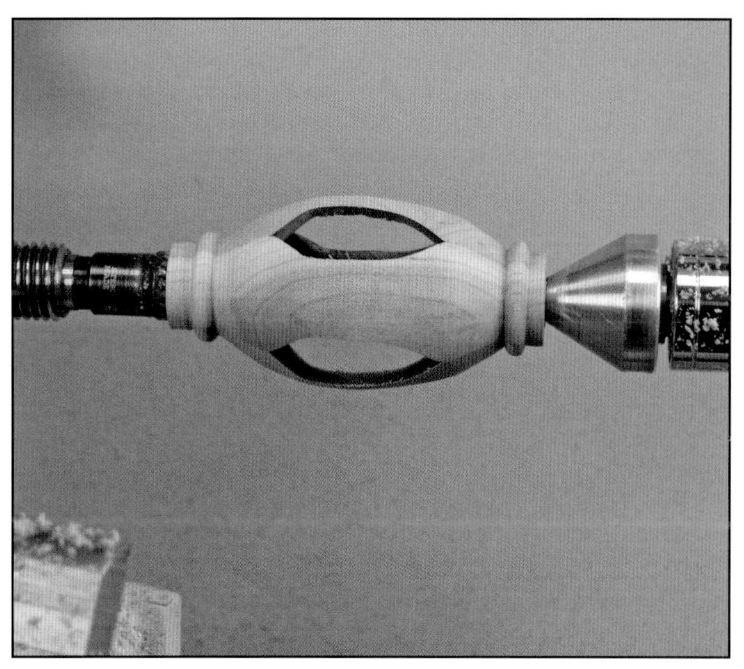
Turn a decorative bead at each end then sand to completion.

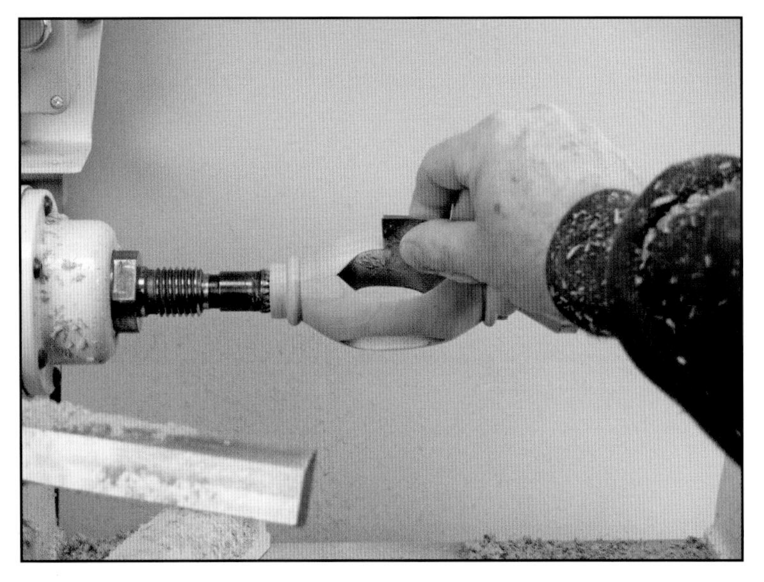

The inside edge must be sanded by hand with the lathe stopped to remove thin fractured fibers.

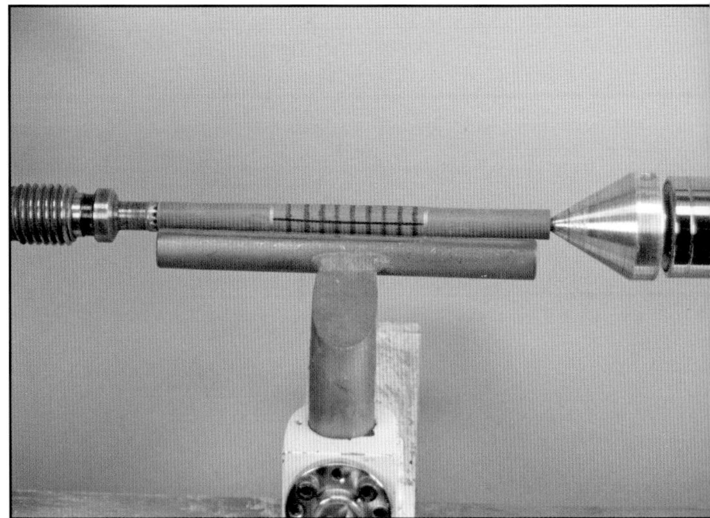

Mark the pitch lines next. As was done for the sliding glass door stop the diameter (1/2-inch) is doubled (1-inch) and the first pitch lines are placed. These are divided in half twice to give pitch lines every 1/4-inch.

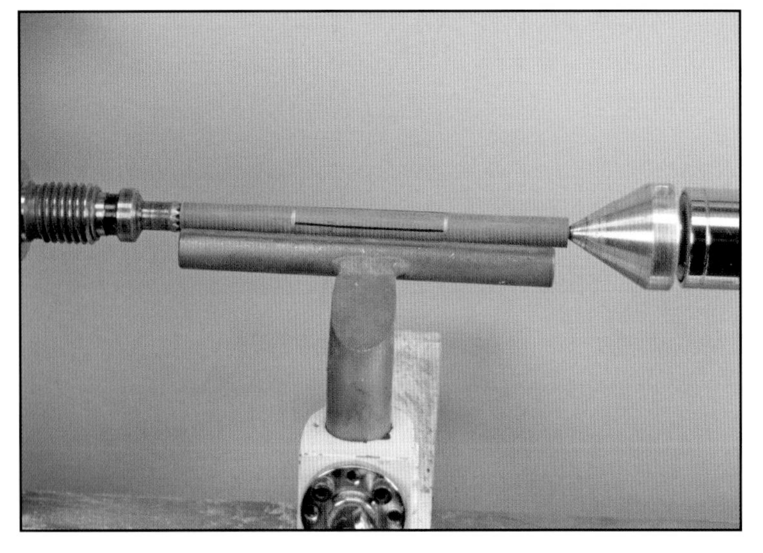

The next step is turning a dowel and making a twist to be placed inside the turning. Pink ivory turned to a 1/2-inch dowel is selected. The silver circumferential marks are defining the area to be seen in the turning and thus the area to be twisted. The remainder of the dowel will be used to glue into the base and top of the stand. The top and bottom are made in the same manner as was done for the plain stand.
Since we are turning a double barley twist we will need 4 start lines drawn in at 6, 12, 18, & 24 on the 24 point index.

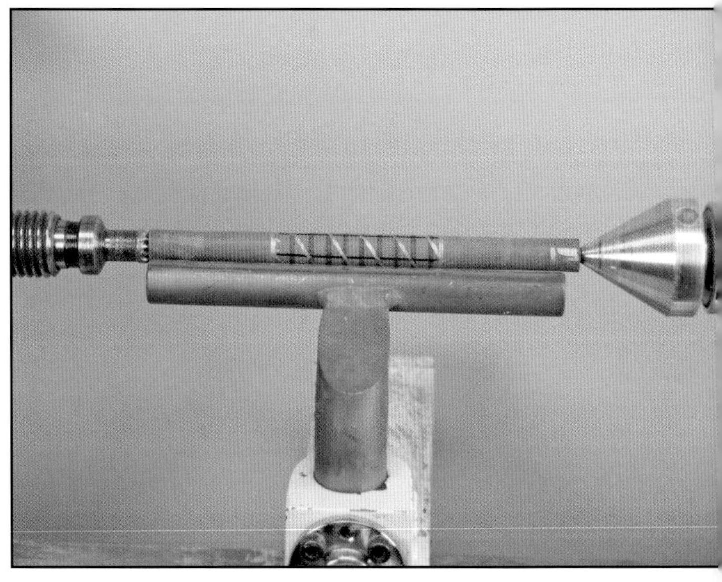

Start at one rectangle at the tailstock and draw a silver line from the lower right hand corner to the upper left hand corner and continue into the next diagonally adjacent rectangle until the headstock end silver mark is reached. Skip a rectangle at the tailstock end and draw a second line around to the headstock silver line. These are the cut lines for the double barley, right-handed twist.

Using a dowel or dovetail saw make scarifications with the blade along both cut lines.

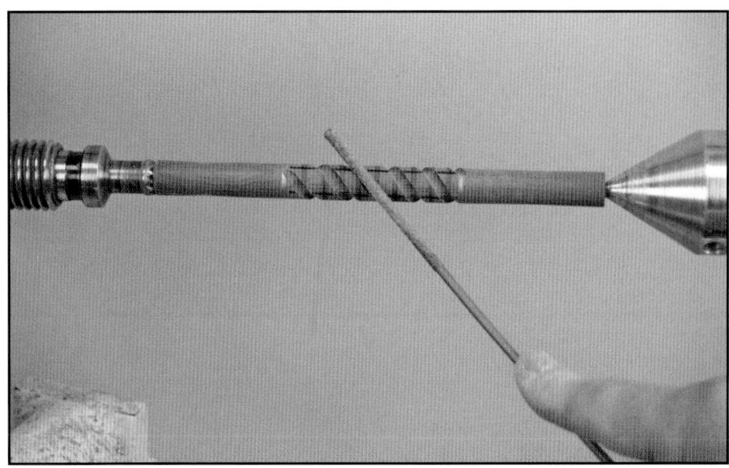

Using the 3/16-inch rasp broaden the coves somewhat.

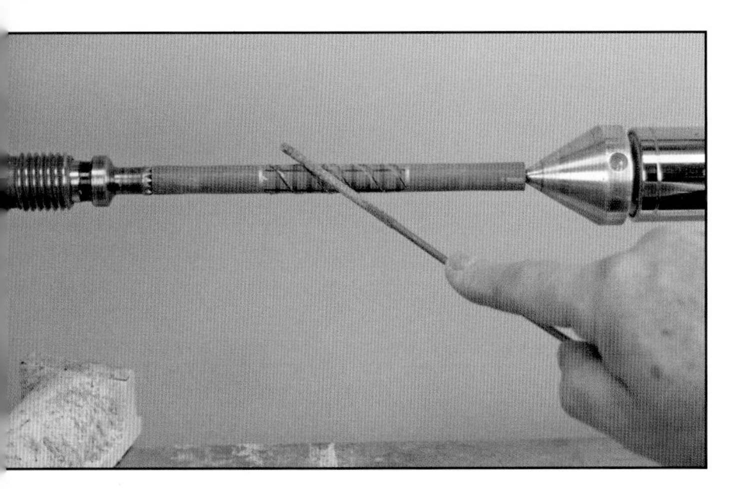

With the tungsten carbide, 1/8-inch rasp, use the scarification marks for guides to begin cutting coves in a parallel fashion.

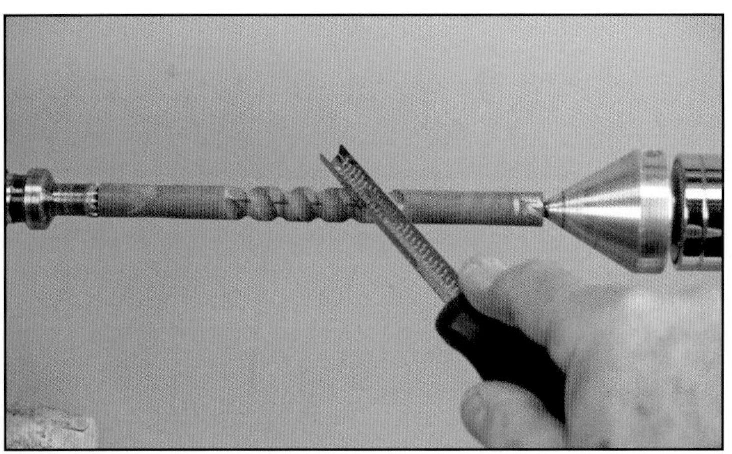

Using a small microplane continue to broaden the coves.

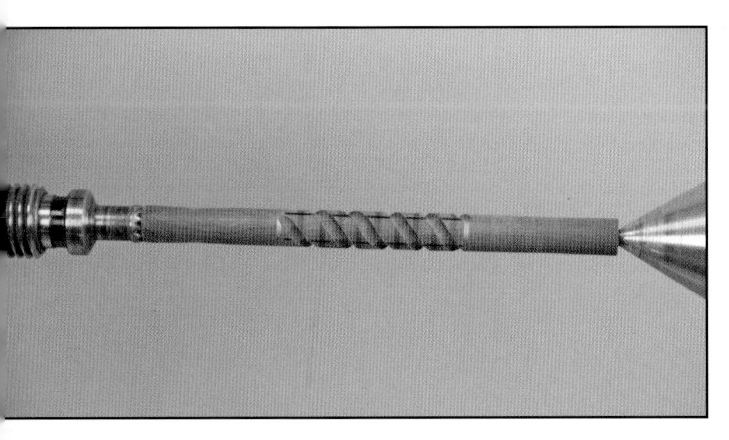

Cut all the way around until nice coves are completed.

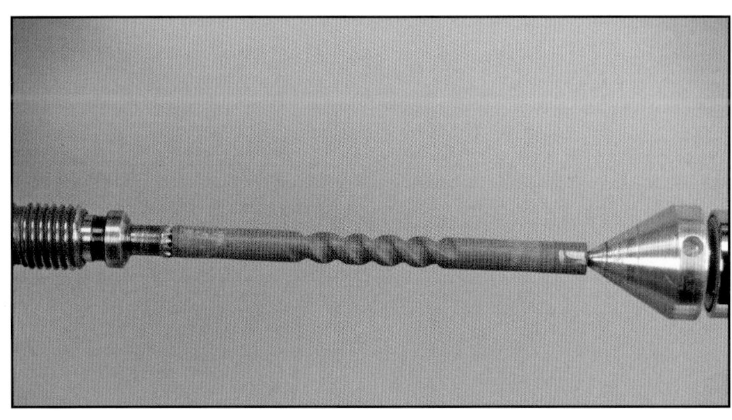

Sand all surfaces with 1/8-inch Vitex paper twisted into ropes (120, 150, 240, & 320 grit) then buff with 0000 steel wool.

The finished cherry plant stand may be used for small potted plants or 6-inch diameter candles.

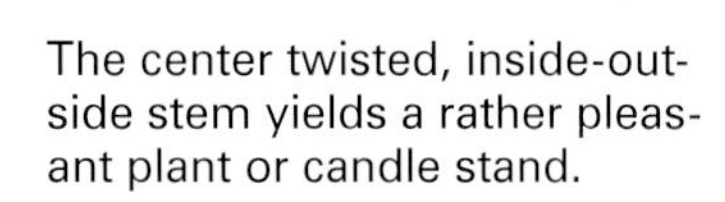

The center twisted, inside-outside stem yields a rather pleasant plant or candle stand.

Chapter 11

Gallery

Door stops made from left over timbers—cherry and Alaska birch—some with spalting.

500 wine bottle stoppers of various shapes, sizes, colors and timbers.

Collection of confetti oil lamps of maple, madrone, Alaska birch, jacaranda, and redwood burls.

Oil lamps of Alaska birch burl, figured myrtle, thuya burl, and pink ivory.

Candlesticks with 5 inch to 9 inch graduated stems of left over cut offs from canted logs of Alaska birch.

Wine caddies of Oregon walnut, Alaska birch, and curly maple glue-ups, some plain and others stained for accents

Collection of various prep plates and small bowls in various timbers.

Hors d'oeuvre plates in cherry and curly maple with purple heart veneer.

Plant stands with stained interiors and twists of blackwood, pink ivory, and yellow heart.

Acknowledgments

First and foremost I would like to thank my wife of 37 years, Susan, for her indulgence in humoring me while I puttered away many hours in the shop tracking dust, dirt, and debris into the house during the late night hours. I should like to take the opportunity to thank Arnie Geiger, as I had in the first book, for his encouragement, source of lumber—both good and bad—in contributing to the production of various pieces in this book, helping make the point that much can be made from little. Once again, I should like to thank Burt Biss not only for his encouragement, legal advice, and proofing of manuscript, pictures, and ideas; but also for the contribution from his scrap wood pile which allowed many portions of the projects described herein to be accomplished. Special thanks go to my editor, Doug Congdon-Martin, whose untiring efforts, Photoshop editing, and infinite patience made both this and the previous book incredibly prodigious and productive endeavors.